I0810599

POSTCARD HISTORY SERIES

Bisbee

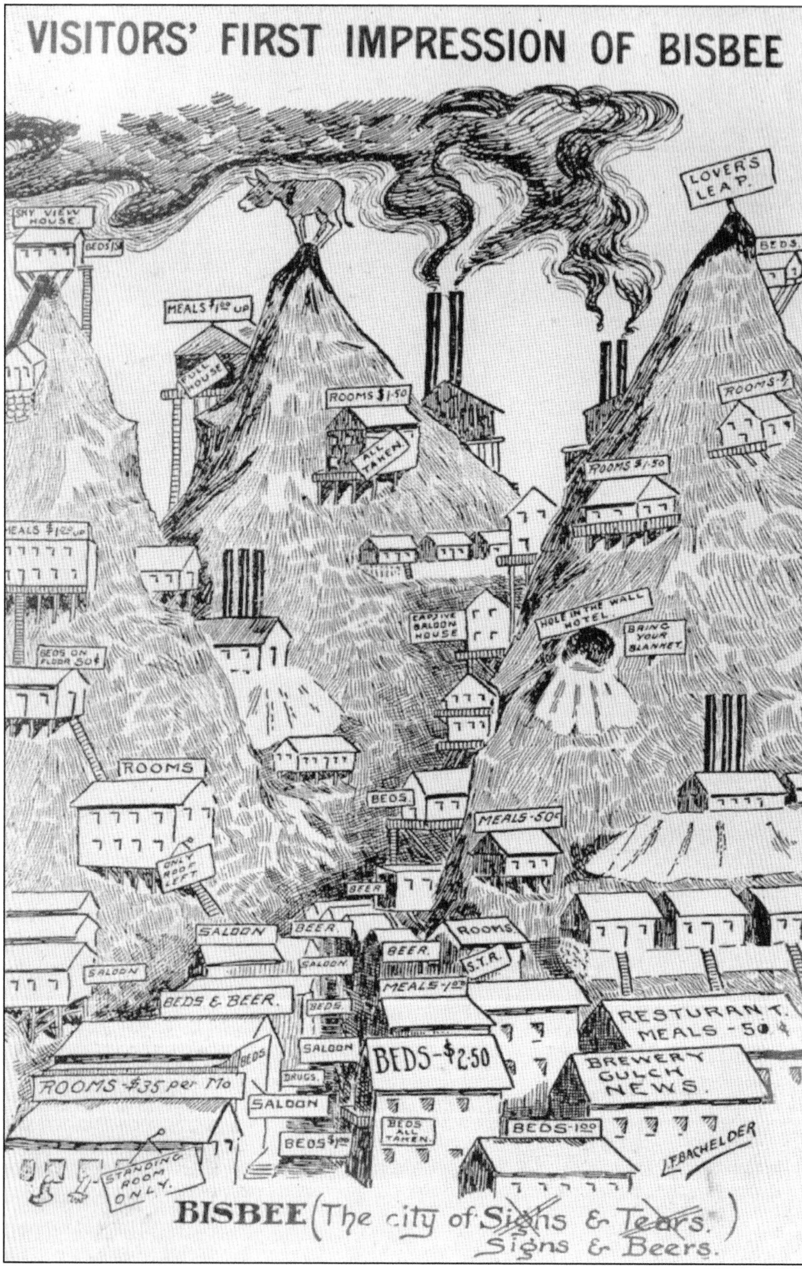

When this illustrated postcard of Bisbee was produced in 1907, the community was the epitome of a bustling mining camp. Smoke filled the air from the thriving mining operations, and the canyon bottoms of the community were filled with saloons, as well as boardinghouses, that climbed up the mountainsides to accommodate thousands of men who worked in the mines. (Courtesy of Graeme-Larkin Collection.)

ON THE COVER: Standing at the location where this 1910s postcard was taken today, one is transported through time, gazing upon the same Main Street that miners strolled down over 100 years ago. The appearance of the thoroughfare has largely remained the same, offering visitors a glimpse into the mining camp's past. (Author's collection.)

POSTCARD HISTORY SERIES

Bisbee

Annie Graeme Larkin

ARCADIA
PUBLISHING

Copyright © 2013 by Annie Graeme Larkin
ISBN 978-0-7385-9996-0

Published by Arcadia Publishing
Charleston, South Carolina

Printed in the United States of America

Library of Congress Control Number: 2012954702

For all general information contact Arcadia Publishing at:
Telephone 843-853-2070
Fax 843-853-0044
E-mail sales@arcadiapublishing.com
For customer service and orders:
Toll-Free 1-888-313-2665

Visit us on the Internet at www.arcadiapublishing.com

To Vance, my parents, siblings, and family,
only with your love and support have I achieved this dream.

Contents

ACKNOWLEDGMENTS

The work that follows has been a vision for over a decade. During this time, numerous individuals have provided their support and guidance. Mere words cannot express how grateful I am to my husband, Vance Larkin. If it were not for his continual support, this book would still be a dream and not a reality. My parents, Nina Graeme and Richard Graeme III, helped in innumerable ways with this work, in addition to teaching me to love and respect Bisbee's history. My siblings, Richard Graeme IV, Doug Graeme, and Jennifer Graeme, were indispensible during this project. Each provided assistance with editing, scanning, and research, as well as giving words of inspiration and support. I am also grateful to my history professors from the University of Arizona, Dr. Barbara Citera and Dr. Jack Mutchler, who always encouraged me to pursue my research about Bisbee's past. A thank-you goes to Mary Walker, who graciously assisted with editing the manuscript. Thanks go to the Bisbee Mining & Historical Museum for the use of three postcards in the book. The staff of the museum, Anna Garcia and Carrie Gustavson, as well as the volunteers, continually offered much appreciated words of encouragement. I would also like to thank the staff of Arcadia Publishing, including Kristie Kelly, Elizabeth Bray, and Alyssa Jones, whose guidance during this process was invaluable. I am also grateful to my friends, as well as the Graeme, Youngs, and Larkin families, who were ready to assist me with the project in any capacity requested. Lastly, I must thank Bisbee. I fell in love with the old mining camp as a child. Countless memories were created as I explored the canyons, admired the beautiful architecture, climbed up hidden staircases, and made lifelong friends. Bisbee has served as the inspiration for both my educational and professional paths and will always have a place in my heart. Unless otherwise noted, all photographs are from the author's family collection.

Introduction

During the 1870s, only the most determined travelers would have dared to venture into the area known today as Bisbee. This remote land was part of the territory the United States acquired from Mexico in the Gadsden Purchase of 1854. Yet decades later, the land remained a desolate region, one largely inhabited by the much-feared Chiricahua Apache. It took a special type of person to explore this bleak desert, one who possessed bravery, combined with the essential skills needed to read and live off of the land. Soldiers and their military scouts were often the type of individuals with these abilities; the accidental discovery of Bisbee's mineral wealth is credited to three men serving in these roles. In the summer of 1877, they were part of a US Cavalry scouting party from Fort Bowie looking for renegade Apaches. Once these men entered the Mule Mountains, the fate of the area was forever changed. The signs of mineral wealth discovered by the scouting party ultimately drew prospectors to the remote mountains, who located claims and later sold them to mining companies. Over time, there were dozens of mining companies in Bisbee, but ultimately, only four would be major players in the camp's mining industry. The Copper Queen Consolidated, Calumet & Arizona, Shattuck-Arizona, and Denn-Arizona mining companies exploited the riches hidden by nature within the limestone hills. However, the buyouts and mergers continued until only one company remained, Phelps Dodge, a descendant of the Copper Queen Mining Company.

Although much of the West was known at the time for its incredible gold and silver deposits, it was copper that would make Bisbee famous. The discovery of copper in Bisbee could not have come at a better time in the context of world history. In the years to come, an enormous demand for electric lighting would be created, and an efficient, inexpensive conductor was needed to carry electricity to illuminate these lights—copper was the ideal metal. Upon learning of the mineral wealth and opportunity for employment, people from all reaches of the globe traveled to Bisbee seeking opportunity. However, this continued to be a journey that the early settlers could not take lightly, as the desert surrounding Bisbee was the territory of some of the most infamous bandits, Apaches, and renegades in the West. With time, the area began to change, as the US military cleared the region of the native peoples and industry moved in.

Realizing a stable workforce was at the core of a successful business, and that skilled miners were scarce, the local mining companies soon built public facilities and brought in amenities to lure families to settle and work in Bisbee. This strategy, coupled with the immense ore deposits, enabled the mining camp to prosper. This once desolate settlement grew to have a library, hospital, stores, schools, and a fine hotel. By 1913, there were more than 20,000 people who lived in Bisbee and the surrounding communities. As mining was the region's economic engine, there were few parts of Bisbee where one could not see a headframe, smell the smoke from the smelter, hear mine bells ringing, or, in later years, feel their home shake from massive open-pit mining blasts.

The original settlement of Bisbee began to spread as more mines developed. Small communities comprised of miners and their families formed around mining properties. A lack of transportation necessitated the men living close to their places of employment in the early years. With time, improvements in the mining camp's infrastructure enabled residents to travel beyond their enclaves to work, shop, and socialize throughout the area.

With thousands of people inhabiting Bisbee and countless individuals visiting this prosperous mining camp, local postcards soon found their way into stores and the hands of individuals. Postcards were a simple and quick way for a person to share a bit of news with their friends and family back home. However, the cards not only enabled one to communicate using the written word, the images on the front also sent an important message to the recipient. One could illustrate the industrial prowess and growth of an area or the devastation caused by a disaster in their community. Also, local mining companies were known to hire photographers to capture their properties to illustrate their industrial and economic might via postcards.

Pictures of the amenities reflecting the quality of life offered in Bisbee helped to lure potential employees with families. Images also acted as a form of communicating recent news, such as labor disputes. Today, sending a postcard of a tragic news event would seem rather odd. However, during a time before people had access to 24-hour news, postcards acted as a means of communicating major events to people who lived hundreds of miles away.

At its peak in the 1910s, Bisbee's population reached between 20,000 to 25,000 people, with the mines at the heart of the city's economy. However, the ores that were once believed to be inexhaustible were depleted by 1975, leaving the community to face life after the mines. Bisbee has now transformed itself from one of the mightiest industrial centers of the West into a community that blends art with its mining history. While Bisbee remains frozen in time architecturally, the function of the businesses within the structures has changed. Once lined with saloons, cigar stands, barbershops, and mercantile establishments, Main Street has transformed into art galleries, restaurants, and antique stores. The unyielding foundation laid by mining in Bisbee proved unbreakable, despite the disappearance of the paternal industry, and has been redefined in the modern age.

One

Arizona's Greatest Copper Camp

When an 1877 US Cavalry scouting party from Fort Bowie entered the Mule Mountains, the fate of the land was forever changed. As Lt. John Rucker, T.D. Byrne, and John Dunn entered the area, they sought not only signs of Apaches but also a suitable supply of drinking water. It was while Dunn investigated a water source that he noticed an outcropping of iron oxide containing a lead mineral called cerrusite. Possessing a fair amount of knowledge of geology, Dunn understood that this was an indicator that other, more desirable, metals might be in the area. Dunn's instincts were correct, and one of the country's most important copper deposits would be exploited from the hills of what is now known as Bisbee.

However, Dunn's instincts were not as accurate when it came to judging the character of George Warren. Warren was a prospector Dunn ran into on his way back to Fort Bowie and to whom he disclosed the discovery of ore in the Mule Mountains. Dunn grubstaked the prospector with the agreement that Warren would return to the area, find additional sites of interest, and include Dunn as one of the locators of the claims. Warren did go to the area and located numerous mining claims, but he left Dunn's name off of each one.

The discovery of copper in the Mule Mountains transformed the land inside and out. Mining companies moved in to extract the red metal from underneath the land, bringing with them jobs for miners and business opportunities for merchants. Mine support buildings, houses, and shops began to sprout up, and the trees that grew on the mountains were cut down to provide lumber for building and fuel for the mines and homes, forever altering the landscape. With time, the once isolated and rugged mining camp turned into one of the most important economic centers in the West.

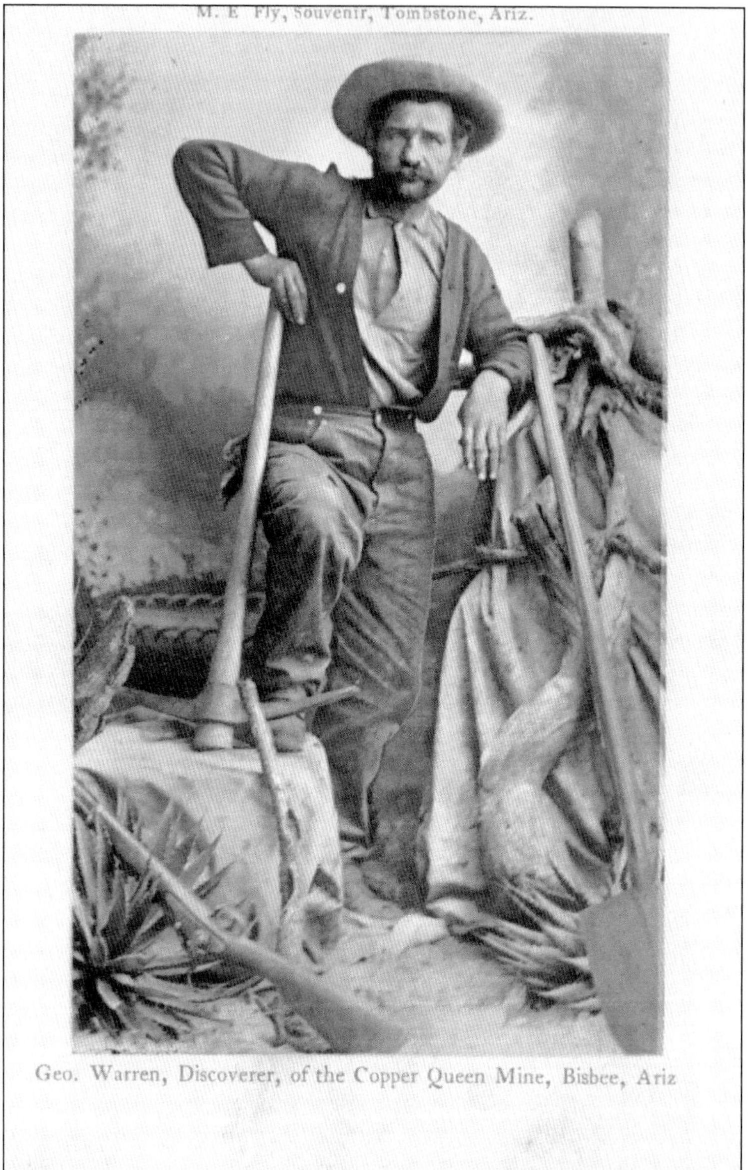

George Warren is one of the most fascinating people in Bisbee's history. When Warren was a child he was taken captive by the Apaches. He was later released when prospectors realized the young boy was being held and traded 15 pounds of sugar for him. It is believed Warren learned about prospecting from the men who freed him. In fact, despite his unscrupulous business dealings with John Dunn, Warren is identified as one of the few individuals who recognized the great mineral wealth in the Mule Mountains. So much so, the area of ore depositions within Bisbee is named the Warren Mining District in his honor. Another distinction bestowed upon Warren was the selection of his likeness from this image to symbolize the miner on the Great Seal of the State of Arizona. Unfortunately, Warren's thirst for liquor and gambling resulted in his losing what would ultimately be millions of dollars in mining claims. Warren died a poor man who cleaned spittoons and swept floors in Bisbee's saloons to earn a little money or get a shot of whiskey.

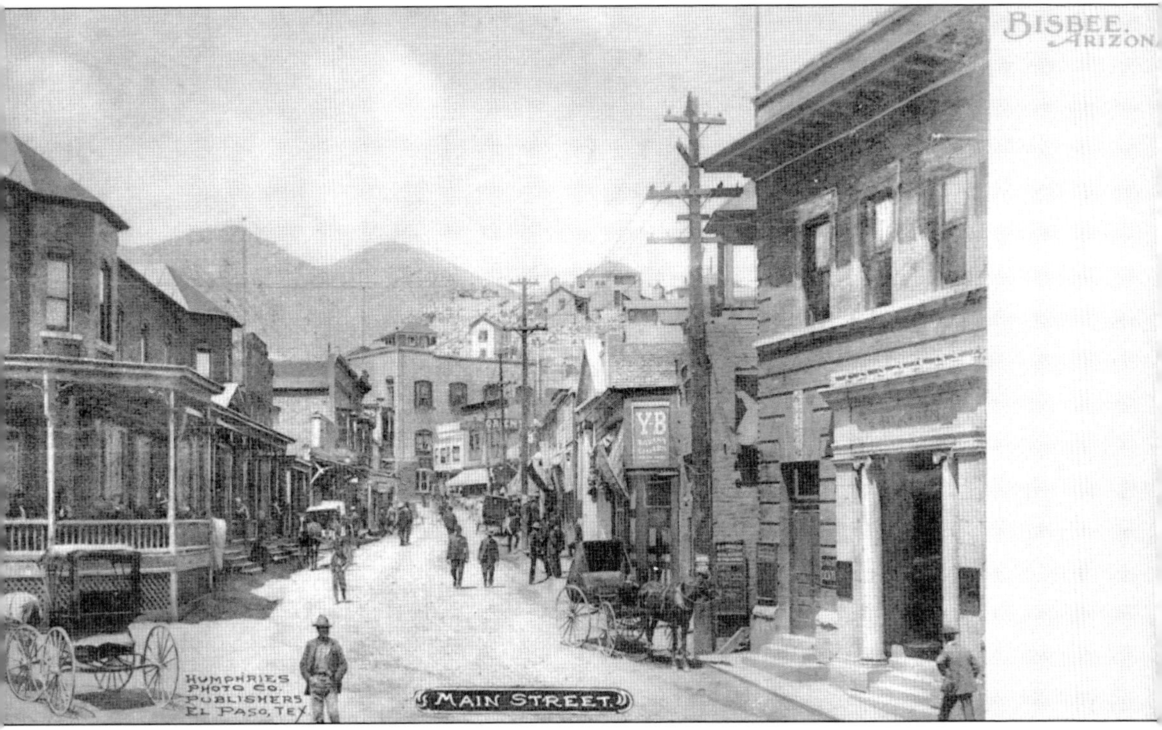

Despite George Warren's lack of good judgment in his business and personal life, his instincts regarding the ore deposits in Bisbee were right on spot. By the time this postcard was released around 1903, the Copper Queen Mining Company alone had extracted over 350 million pounds of copper from Bisbee, and the ore seemed limitless. The *Bisbee Daily Review* boasted that Bisbee was the "most promising, greatest, copper-producing camp on the face of the earth." It was "a place for the man who will work" and "a place for a man who will take a chance of being a workman today, [and be] a wealthy man tomorrow." People held faith in the future of the mining camp, and this was visible in Bisbee's postcards. This image illustrates the community's confidence through the development of Main Street. Many of the buildings shown here were constructed using the lasting materials of brick and mortar.

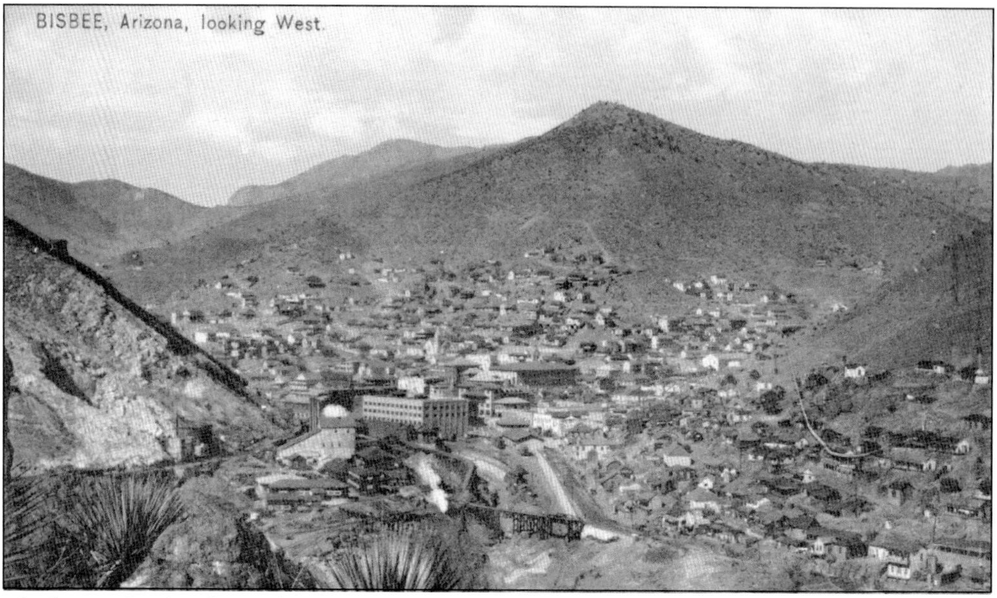

The heart of Bisbee lies at the intersection of Tombstone Canyon on the left and Brewery Gulch on the right in this 1914 postcard. How the community grew was based largely in part on the locations of the local mine sites. This is shown here, with the Czar Mine and smelter at left center and the community sprawling out from the industrial area.

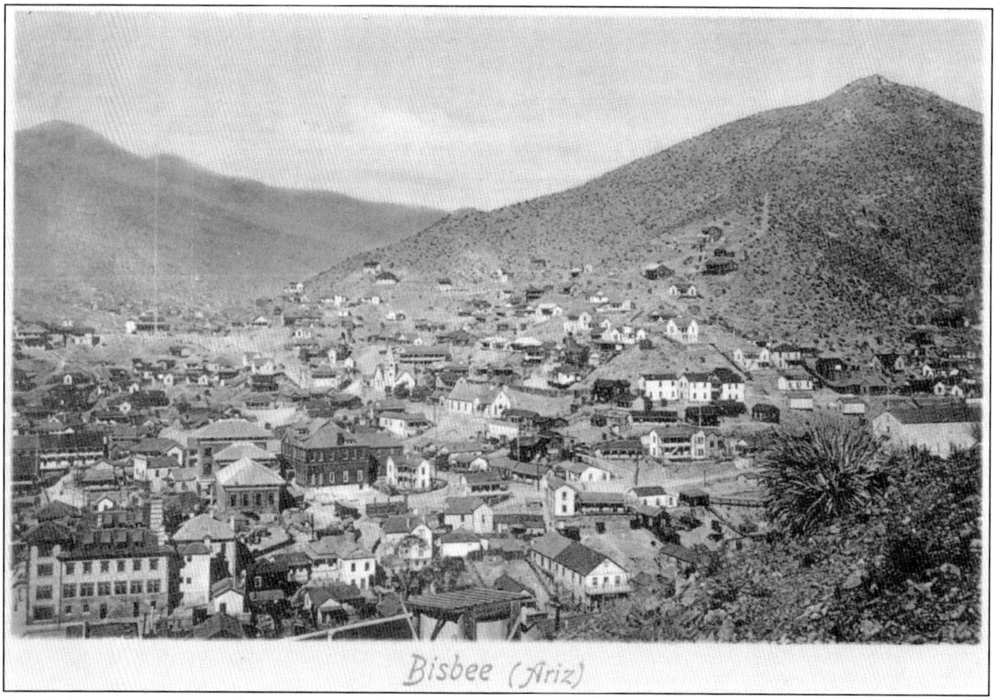

Both the pace of population growth and the geography of Bisbee resulted in the community growing in a unique manner. When space became limited on the canyon floors, dwellings began to creep up the hillsides. The hills were also sought to build homes as they were typically protected from the frequent flooding caused by the torrential monsoon rains.

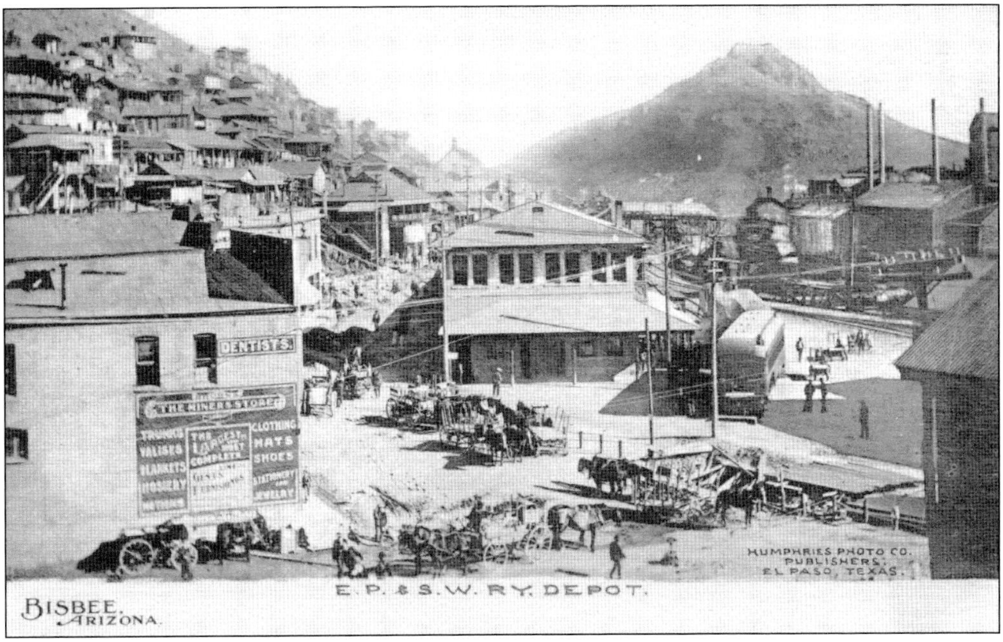

The remote location of Bisbee presented transportation challenges for the mining companies. In order to lower costs and achieve maximum profits, the companies brought the railroad into the mining camp. The rail line, with its depot shown here at center around 1902, entered the community from the east, going around the Mule Mountains to avoid the steep climb up the terrain.

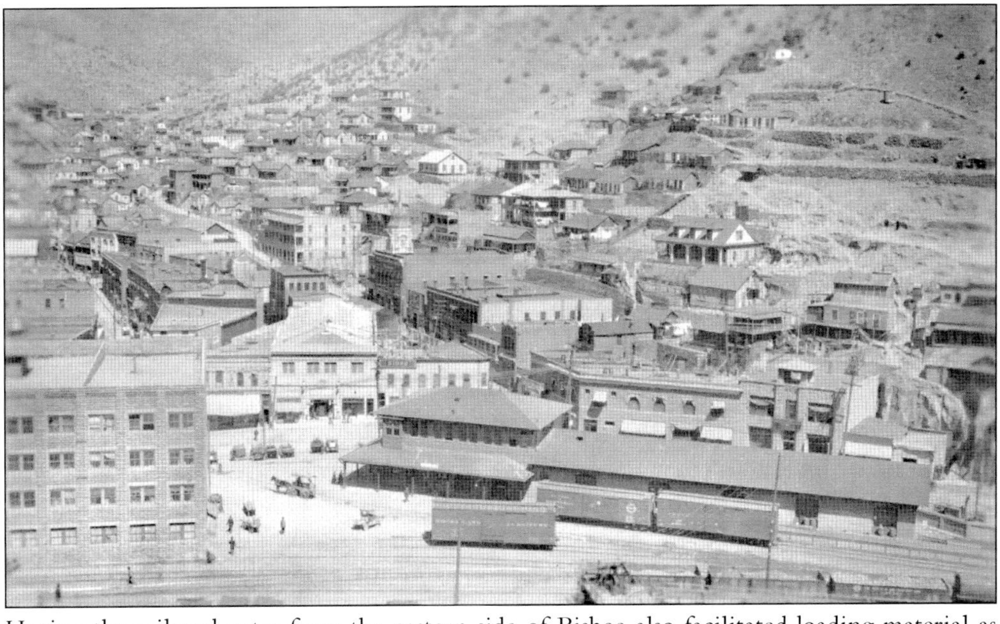

Having the railroad enter from the eastern side of Bisbee also facilitated loading material as it passed by many mines; the cars would be loaded as they went by the operations. The right foreground of this image shows railcars filled with ore, ready to transport the community's copper to the metals market.

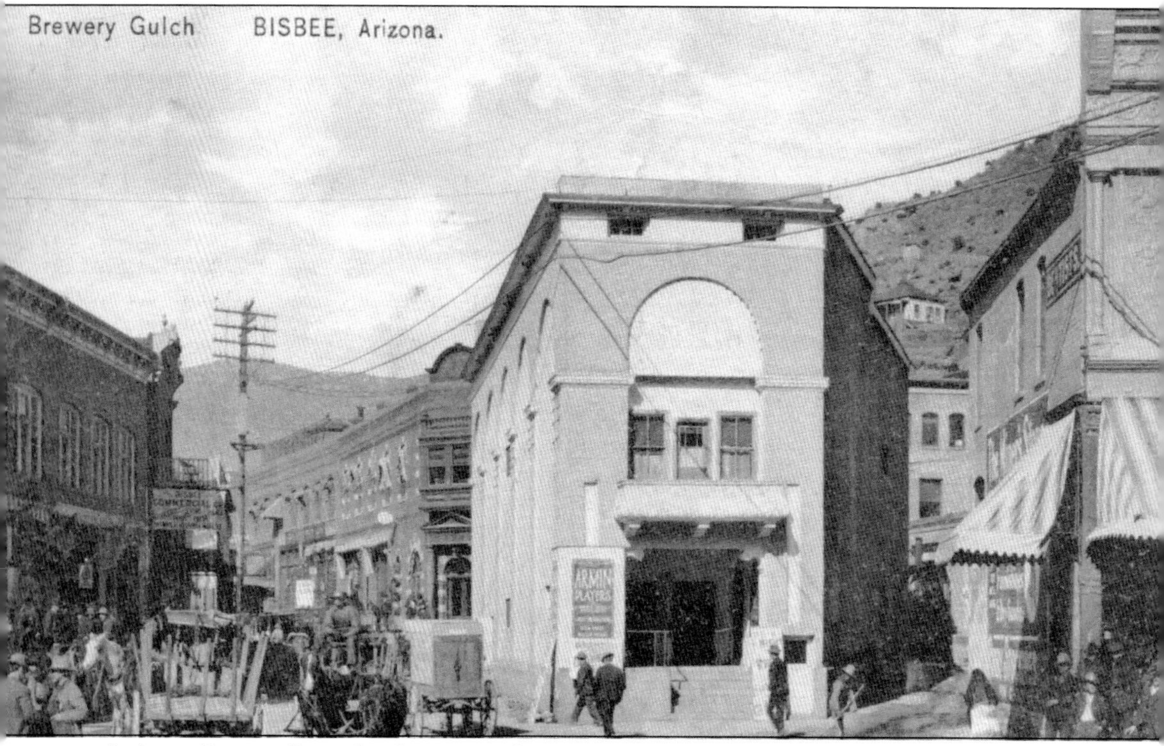

A short distance from the depot was the entrance to Brewery Gulch, captured here around 1910. It was a vibrant neighborhood, and postcards of it capture this excitement. Historian Joe Chisholm described the significance of Brewery Gulch as being "to the Mule Mountains of Arizona what the Appian Way was to Rome, what Broadway is to Manhattan Island." Many thirsty miners called the rust-colored hills of Bisbee home, and eager entrepreneurs moved into the community to start businesses that accommodated the essential, as well as sordid, needs of the many men in this bustling mining camp. Although the entrance to Brewery Gulch had the appearance of an average business district, this was merely a facade for what was the most notorious neighborhood in the city.

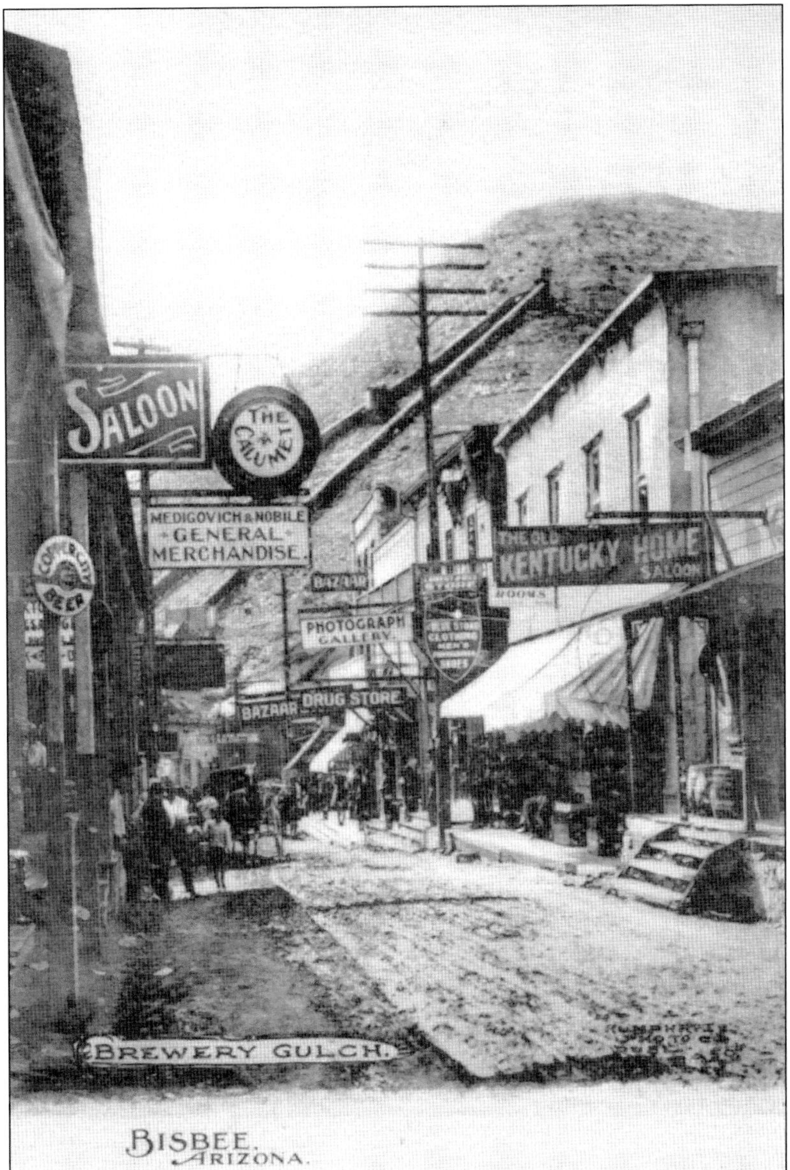

A 1900 article stated that Bisbee "has as many gambling houses, liquor saloons and brothels as it needs—and probably more." Many of these businesses were located in the infamous Brewery Gulch, which housed an estimated 20-plus drinking and gambling establishments, as well as the red-light district. Brewery Gulch was clearly a neighborhood geared to meet the various needs of men of the mining camp. While the city offered many extracurricular activities for its residents, the reason for Bisbee's existence was never far from view. Above the buildings are the long smokestacks from the mine's smelter climbing up the hill. While having the smelter located in Bisbee caused a great deal of air pollution, people often found comfort in the smoke. As long as smoke was billowing from the stacks, there would be a job to return to tomorrow.

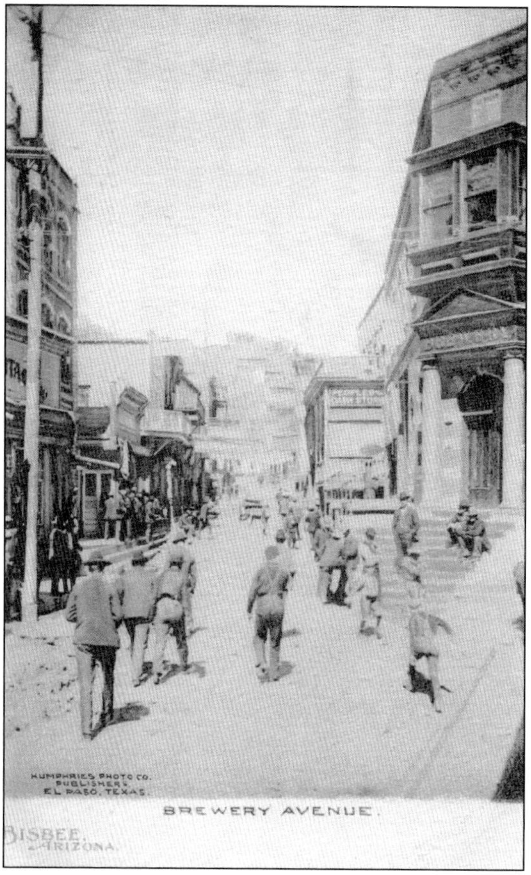

While Bisbee had numerous opportunities for one to partake in lurid activities, it tried to maintain a certain level of decorum by establishing clear boundaries for the females of the mining camp. The first ordinance passed by the Bisbee City Council banned women from entering "saloons in the town of Bisbee." However, due to a loophole in the law, women were allowed in those types of establishments situated in the designated red-light district at the north end of Brewery Gulch. A 1908 map of this area reveals that 36 locations were utilized by ladies of the evening. Although it was not a secret that the red-light district existed in Bisbee, one would be hard-pressed to find a postcard that exposed this part of the mining camp or the women who worked there.

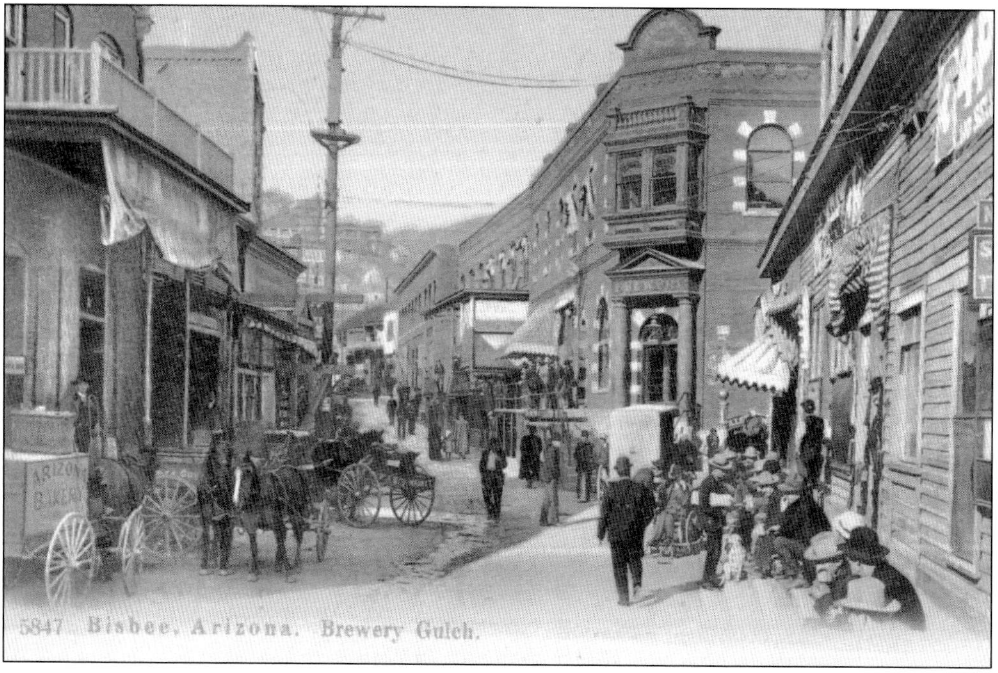

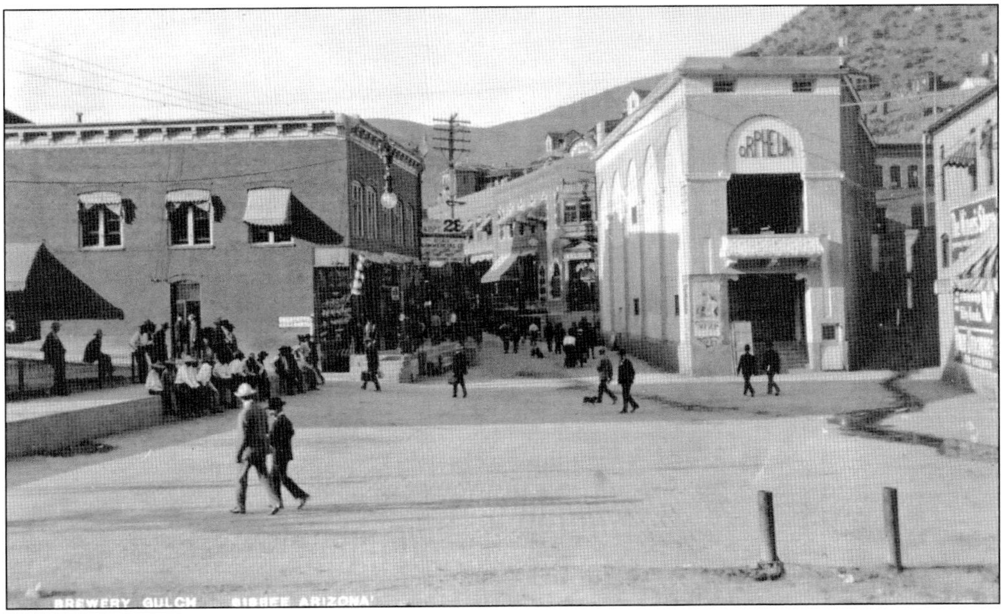

The entrance to Brewery Gulch was home to the Orpheum Theatre, pictured here around 1908. It was one of the cultural amenities Bisbee offered its residents. In addition to being known as a premier copper camp, Bisbee was also considered "the best show town in the territory." Having such a reputation to maintain, the Orpheum strove to bring the best shows available to Bisbee.

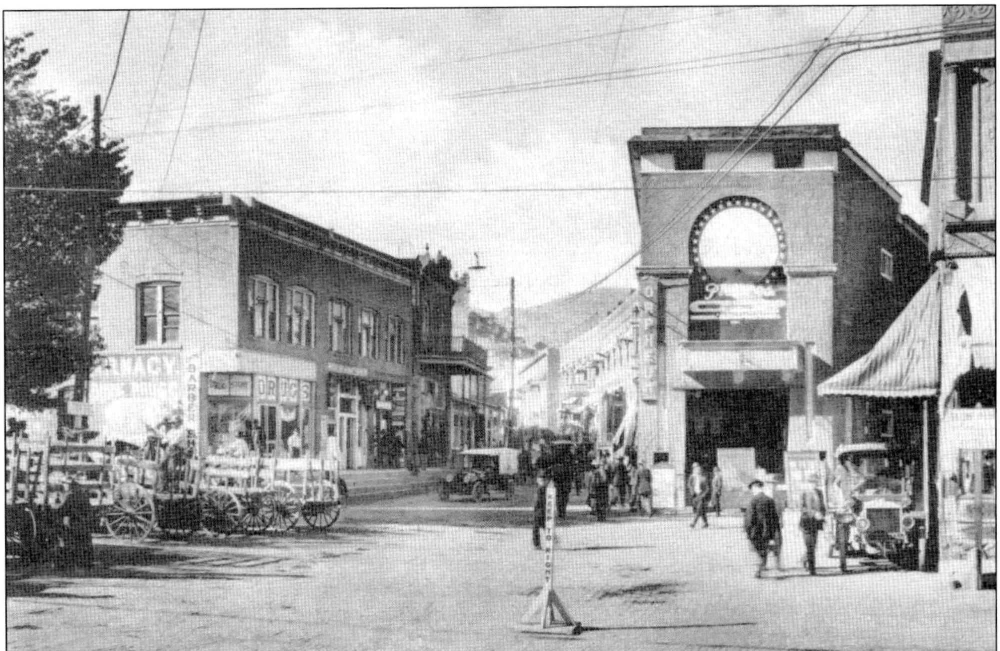

The Orpheum presented plays, as well as motion pictures, with the popularity of films becoming evident when the theater's sign changed to read Paramount Pictures around 1916. The dawning of a new era in transportation is also visible in this postcard, as the transition from the horse to the automobile is apparent.

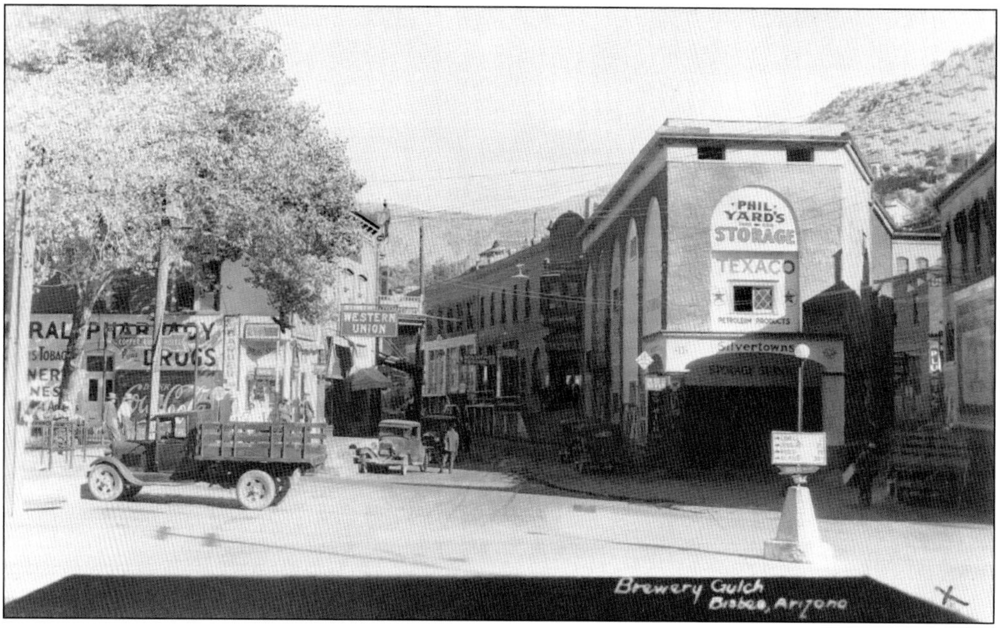

As time passed, the entrance to Brewery Gulch reflected the needs of a changing society, and the function of the building that once housed the Orpheum Theatre was altered. The advent of the car created a need for automobile-related shops. The building became the home of Phil Yard's, shown above around 1930, and then Goar's Garages in the late 1930s. The street beyond the entrance also changed as time passed. What one postcard sender described as the once "wild & woolly district" was faced with laws forbidding gambling and liquor, two of the foundations that gave Brewery Gulch its infamous reputation.

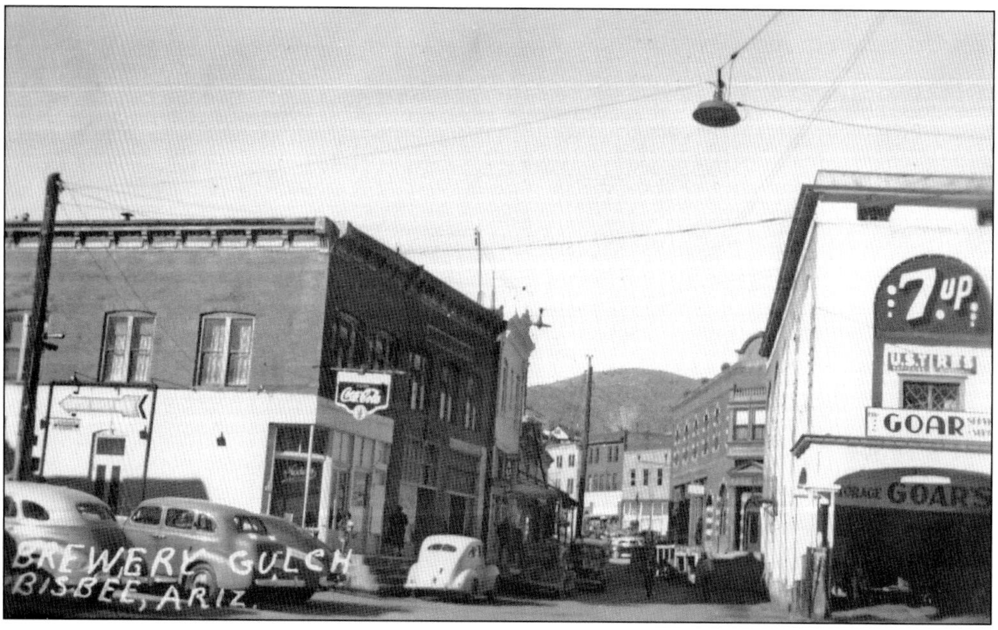

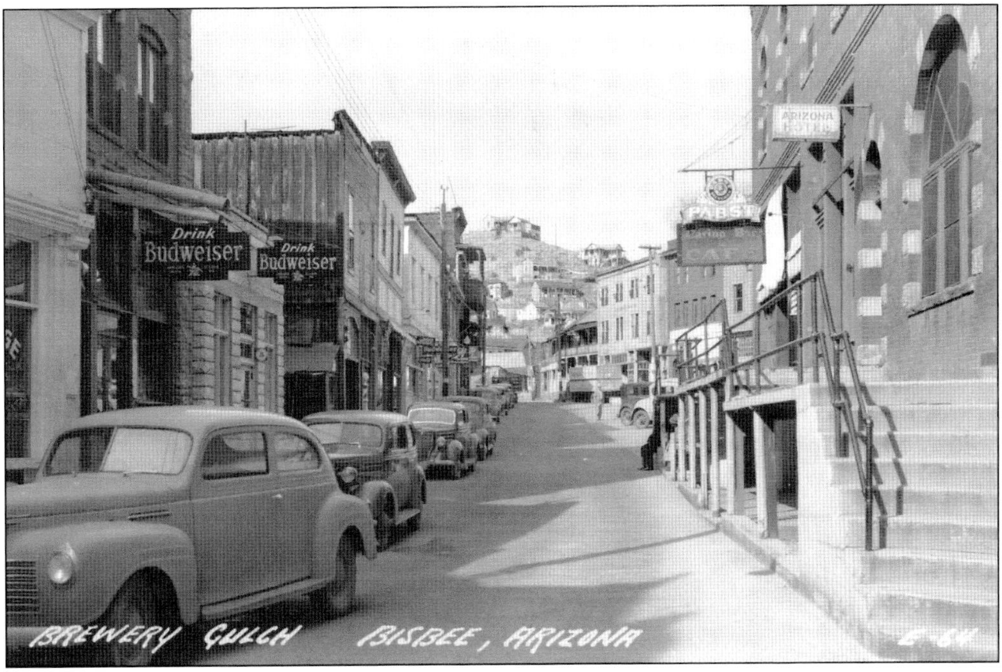

Once Prohibition was repealed, Brewery Gulch again became the home to local saloons. However, the street never returned to the glory days of being the most notorious street in this frontier mining town.

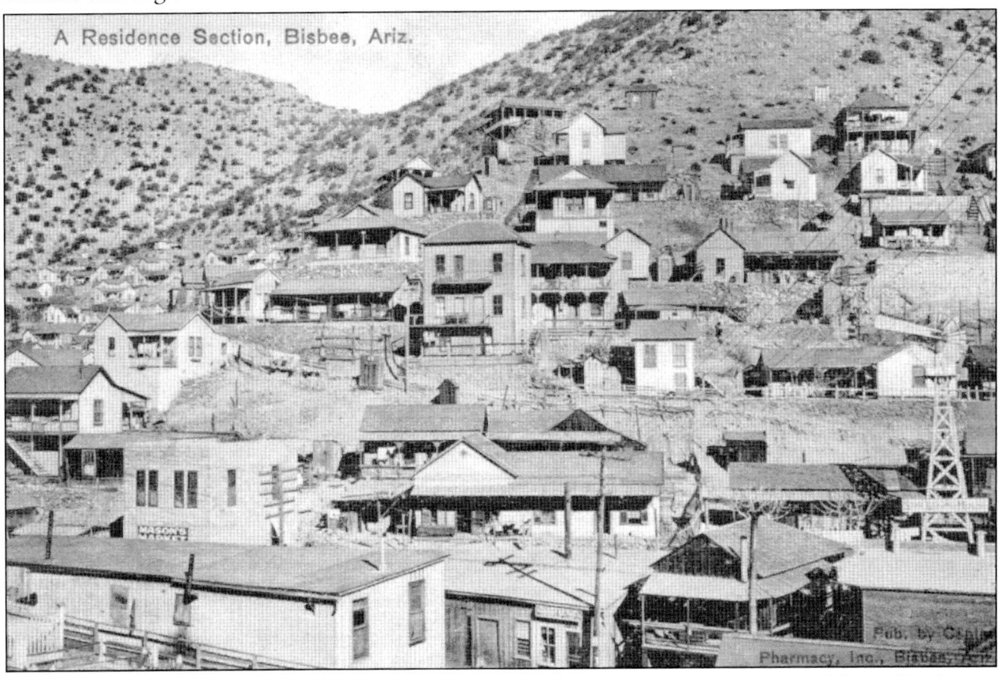

While many businesses occupied the bottom of Brewery Gulch, home and boardinghouse owners looked to the hills to build their structures. This 1910 image captured the canyon bottom at Brewery Gulch, with structures working their way up the hill to OK Street. Bisbeeites built an amazing eight tiers of houses that climb up the steep hill.

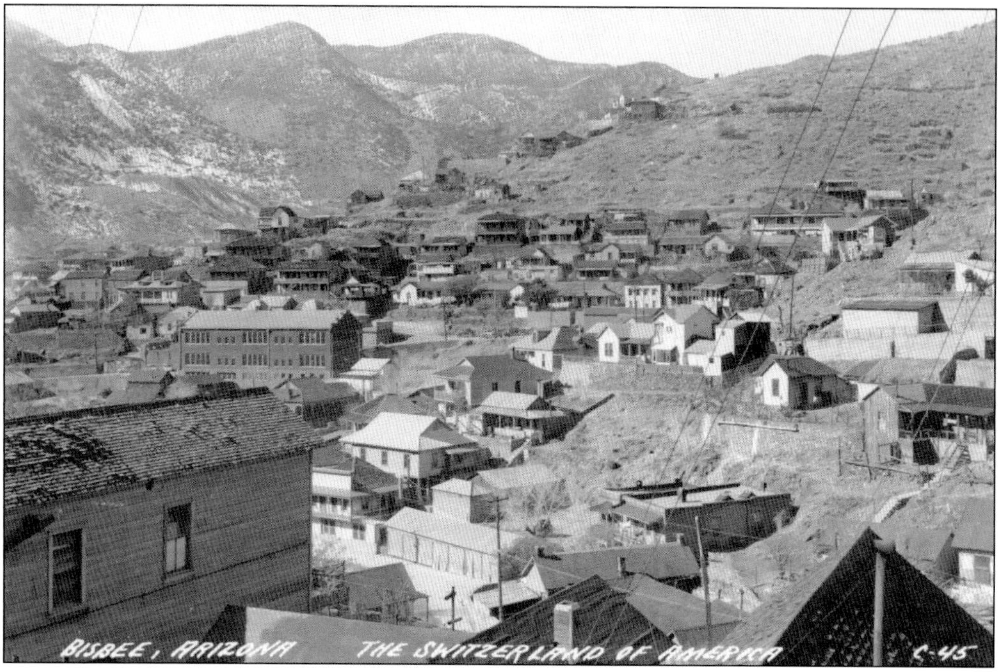

The building techniques employed in this mountainous community were often topics of discussion for those sending postcards. People were amazed at the construction of buildings in canyons and on hillsides. One person declared, "There is not one level spot in this mining town!" The look of the city also resulted in comparisons to communities in Europe, such as the card above that declared Bisbee "The Switzerland of America." Moreover, the language chosen to describe the city is never ordinary. Writers frequently used words such as "interesting," "strange," and "unique" to explain Bisbee's landscape.

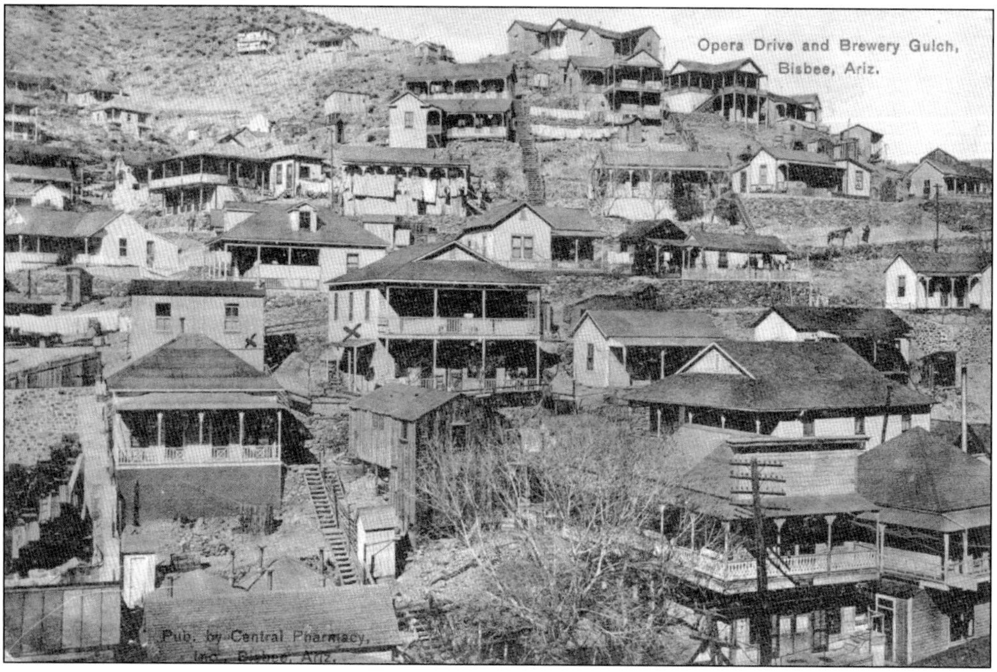

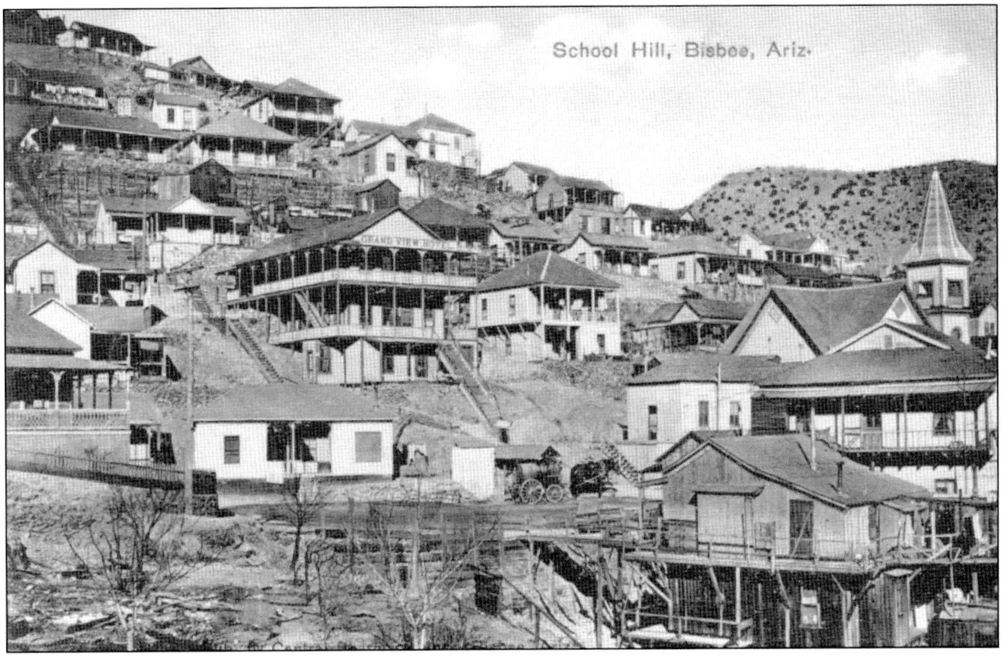

Countless homes in Bisbee were built on such steep inclines that residents often had to build stairs leading to their houses. This is visible on the left side of the 1905 image above. Having one's home so high up a hill presented potential challenges for some residents. One postcard reads, "Some of these homes are 400–500 [feet] from the street level. A man had better stay sober when he goes home or he's in for a good fall."

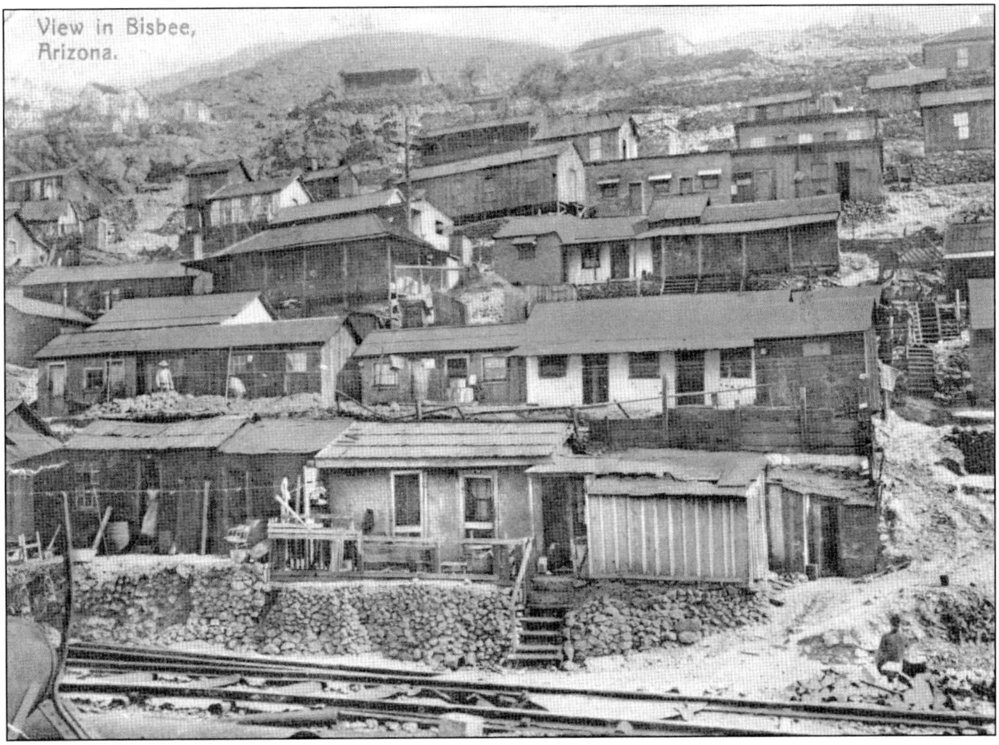

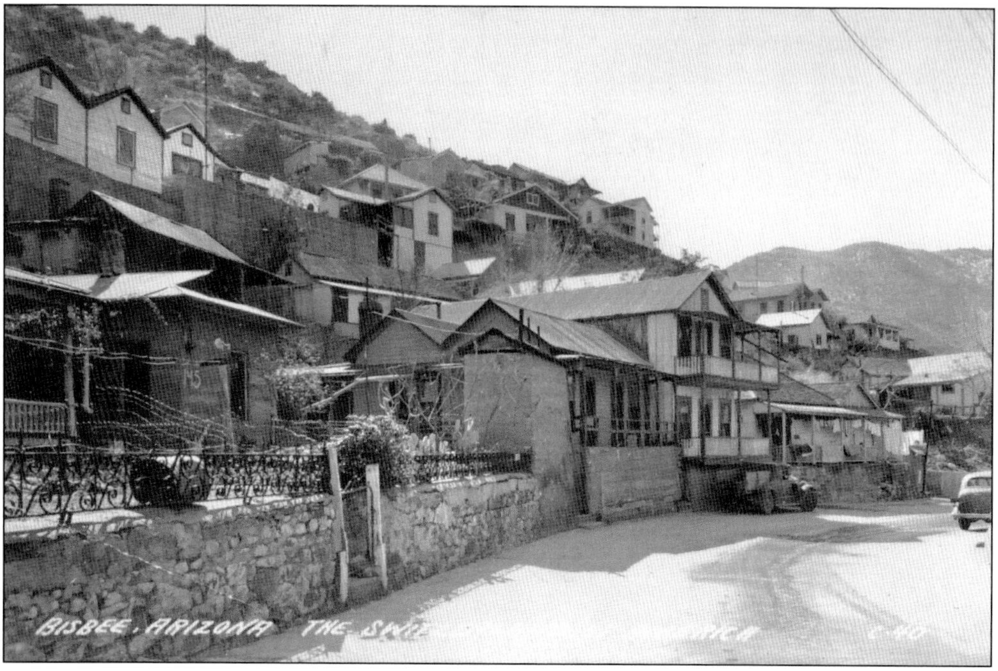

Many dwellings on the canyon bottoms in Bisbee possessed high walls, as is evident in this 1930s image. Bisbee was known to receive heavy rains during the summer months, which often resulted in flooding. These high walls helped prevent the floodwaters from entering homes and businesses.

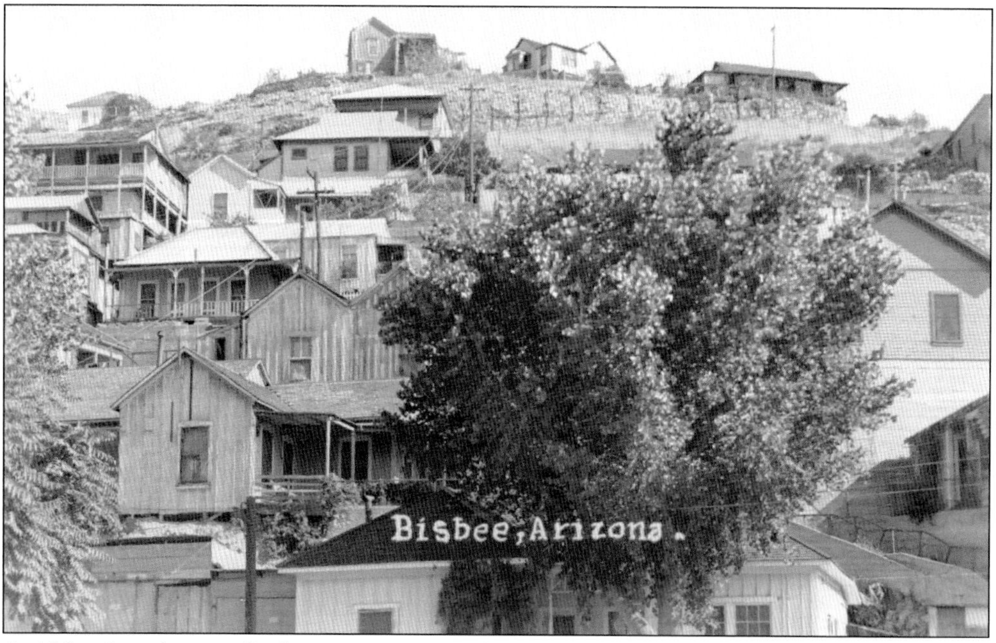

As if floods and steep climbs were not enough for residents to contend with, Mother Nature tested some Bisbeeites in other ways. There were at least two occurrences of homes being destroyed from boulders rolling down the surrounding steep hills. Yet despite the terrain's unusual challenges, the community continued to grow along the hills and the canyon floors.

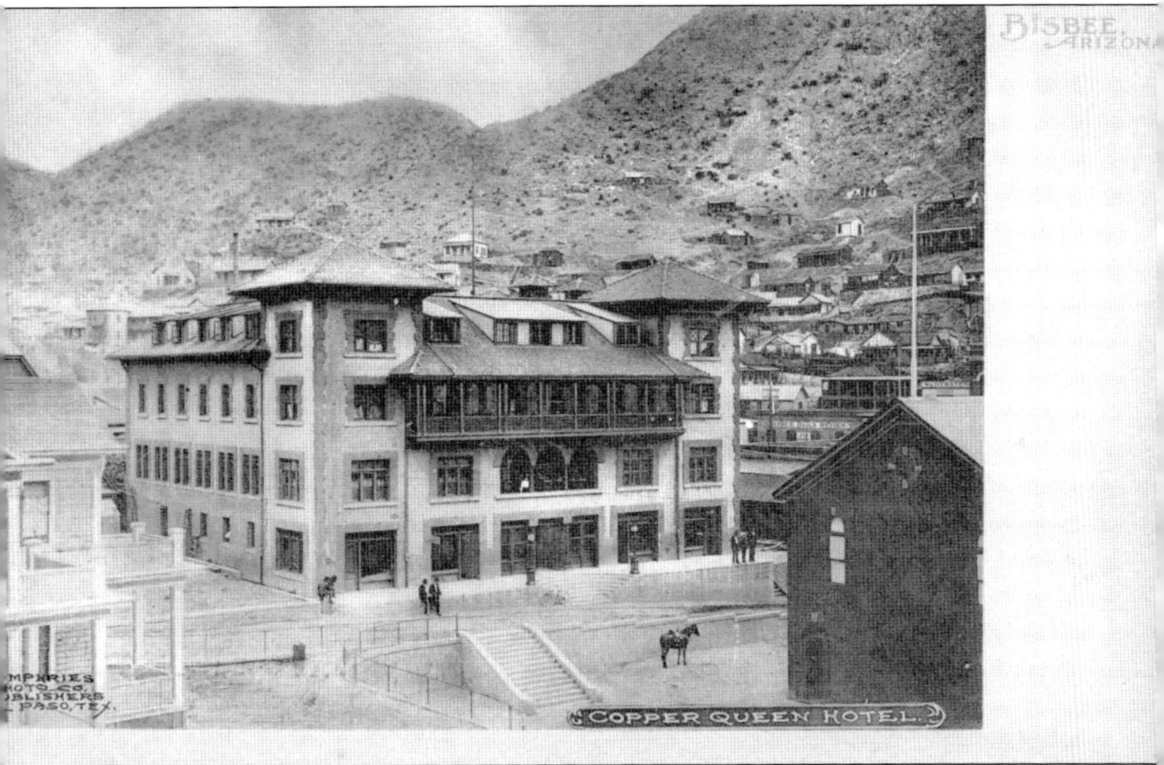

Not far from Brewery Gulch is the Copper Queen Hotel, whose name was "once suggestive of the great industry from which it takes its name." The Mission-style hotel opened in February 1902 and has earned the status of being the oldest continually operating hotel in Arizona. The hotel was built in response to the mining company's need to have a suitable place to house visiting business associates. When it opened, the *Bisbee Daily Review* called the lodging facility an "indispensible institution that meets the demands of a much felt need." It was a first-class facility complete with dining and billiard rooms, a kitchen with the most up-to-date appliances, and even the finest carpet and draperies from Marshal Field & Co. in Chicago.

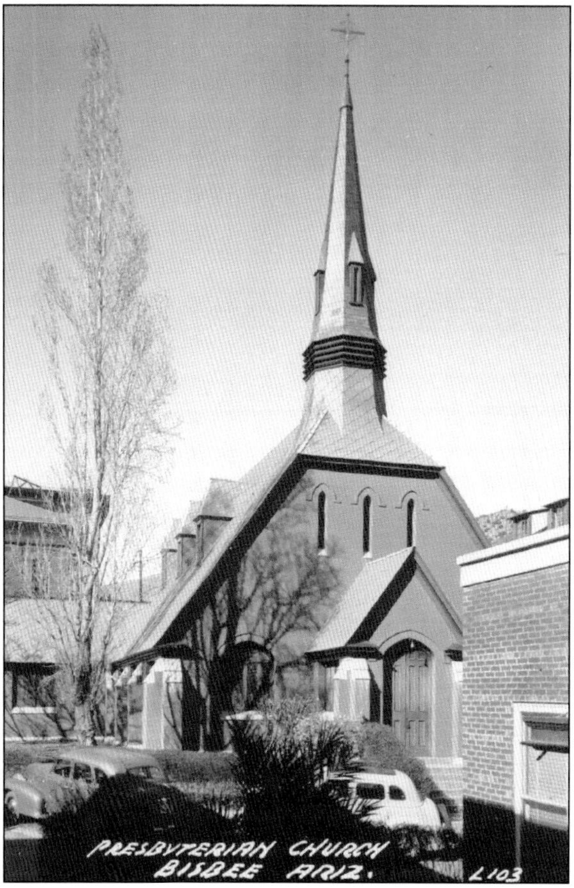

The Presbyterian church was built next door to the Copper Queen Hotel and was one of Bisbee's many houses of worship. Members of the Phelps Dodge Company realized the residents of Bisbee had a variety of desires that they needed to have met, including those of a spiritual nature. The company assisted in erecting the church in 1902, and it continues to maintain a congregation today.

Bisbee was best known for its rowdy extracurricular activities, but there were opportunities for the city's male residents to partake in clean entertainment. This included the Young Men's Christian Association, pictured here about 1940, where men could indulge in respectable recreational and social activities. Although individuals were skeptical of the wholesome organization's success in a Western mining camp, it acquired over 600 members in merely 10 months.

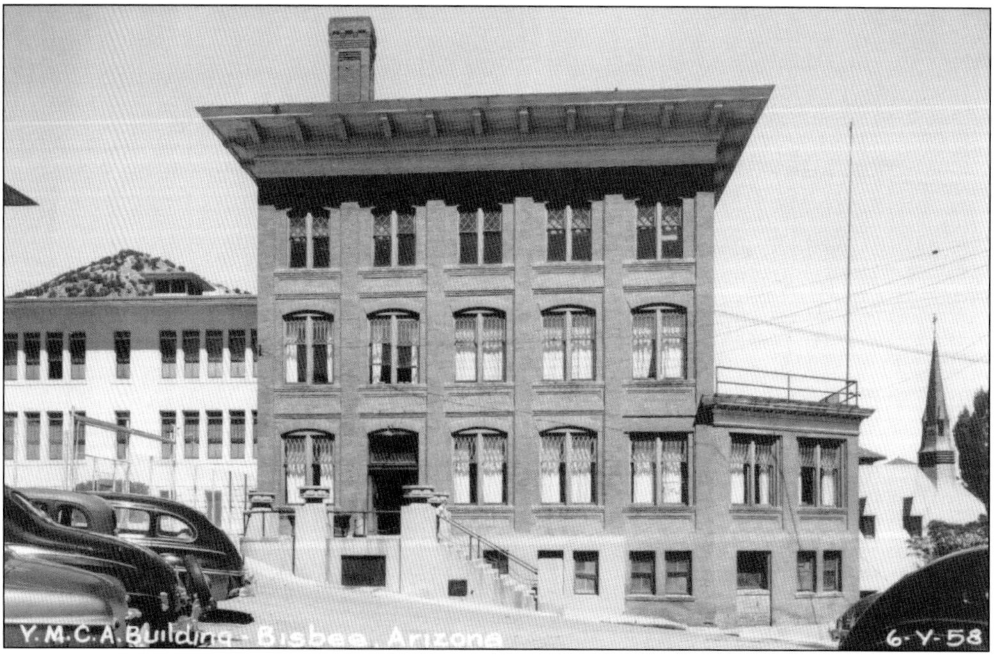

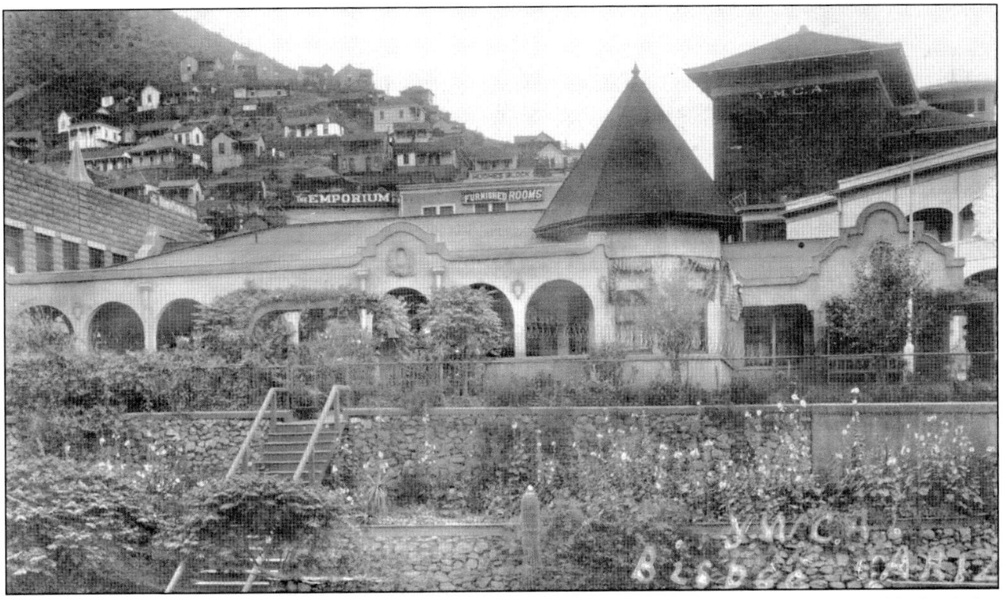

Grace Dodge, a member of the Dodge family of the Phelps Dodge Company, established the Young Women's Christian Association in Bisbee in 1908. With the assistance of Dodge and the mining company, the first Young Women's Christian Association was housed in the former home of mining executive Walter Douglas. The *Bisbee Daily Review* noted the organization's structure offered space to accommodate the male callers of women. The typical housing situation for young females in the mining camp forced them to visit with their male friends in their rooms. The Young Women's Christian Association offered a socially acceptable environment that was not "demoralizing and degrading" for men and women during a courtship. In a few years, another building was constructed on nearby Howell Avenue to house the organization, shown below, about 1940.

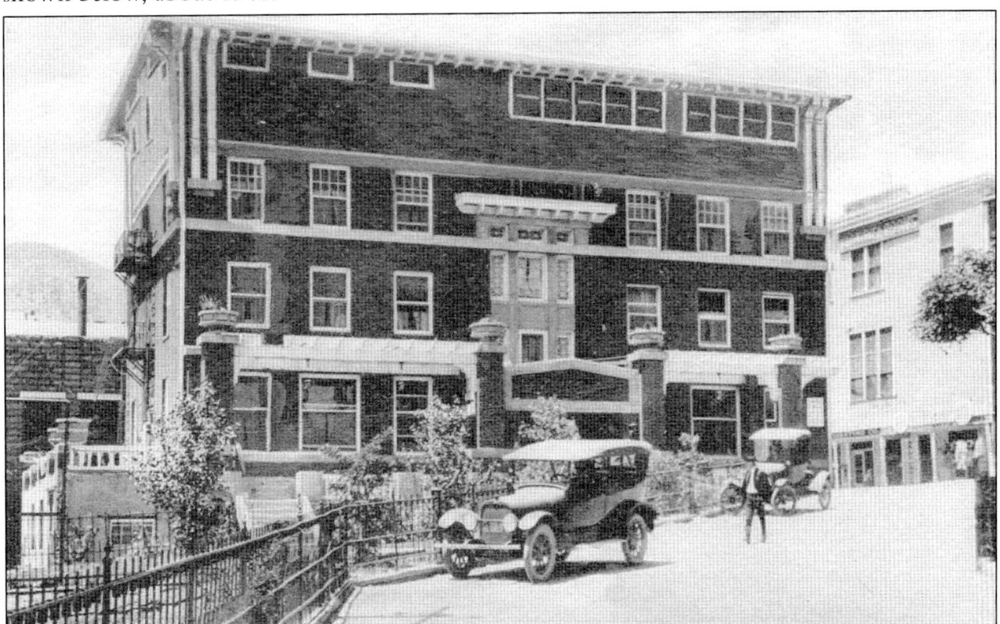

Y. W. C. A. BISBEE, ARIZ.

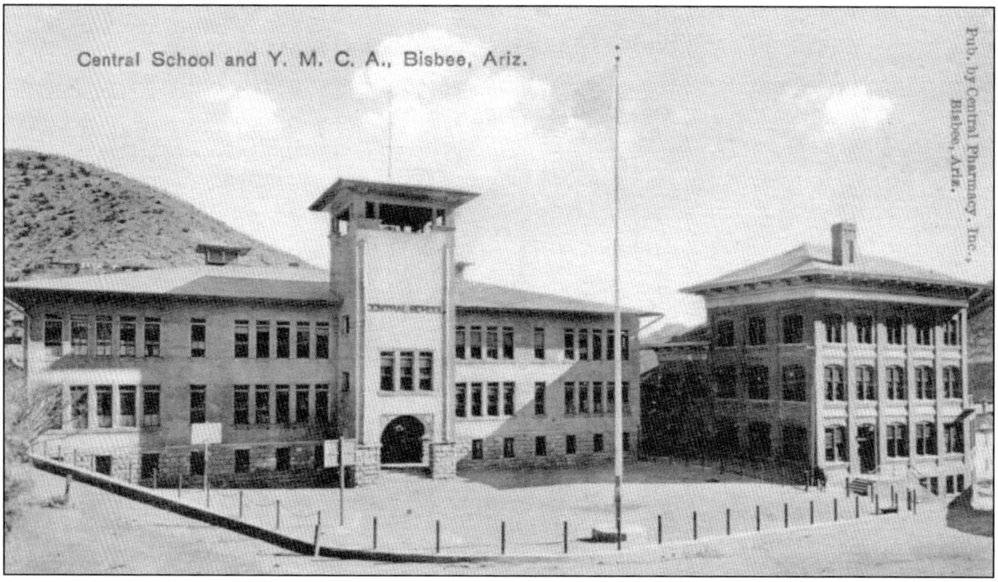

As the community grew, so did the need for a public educational system. The first schoolhouse in Bisbee appeared in 1881; two years later, the Copper Queen Mining Company built an adobe school for the community. The latter building witnessed many renovations over the years, as rooms were added when the number of students increased. However, by 1905, the school's resources were stretched to their limits and a needed reorganization of the educational system took place. This restructuring accommodated all grade levels, and buildings were erected to hold the increasing number of students. This includes Central School, which was built upon the site of the old adobe school. Bisbeeites were proud of the city's educational system, with the local newspaper stating it stood as a model for other frontier communities to look toward.

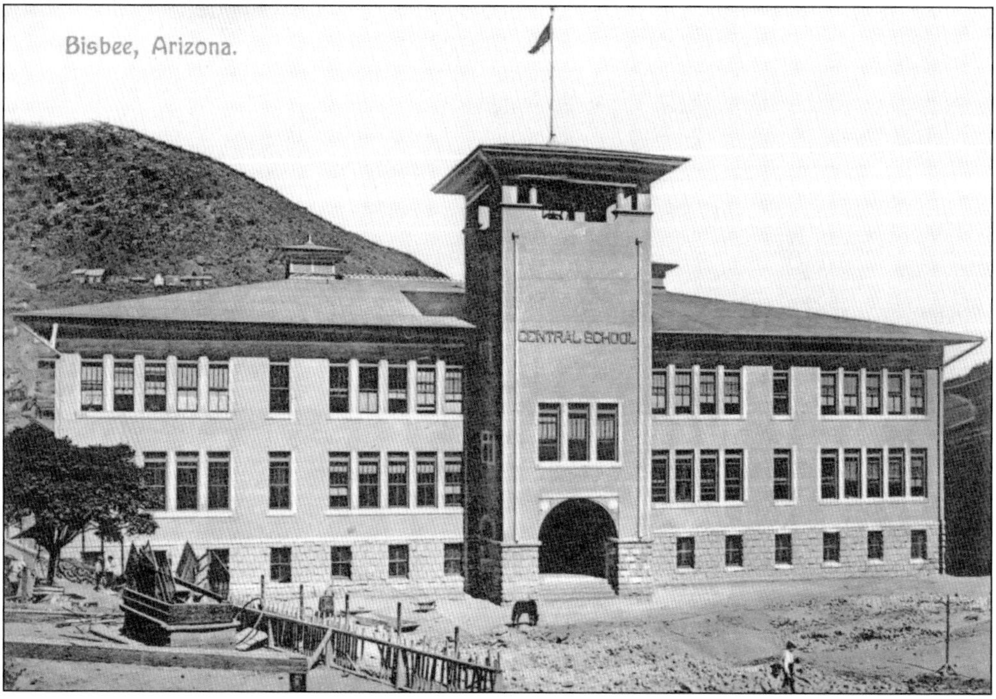

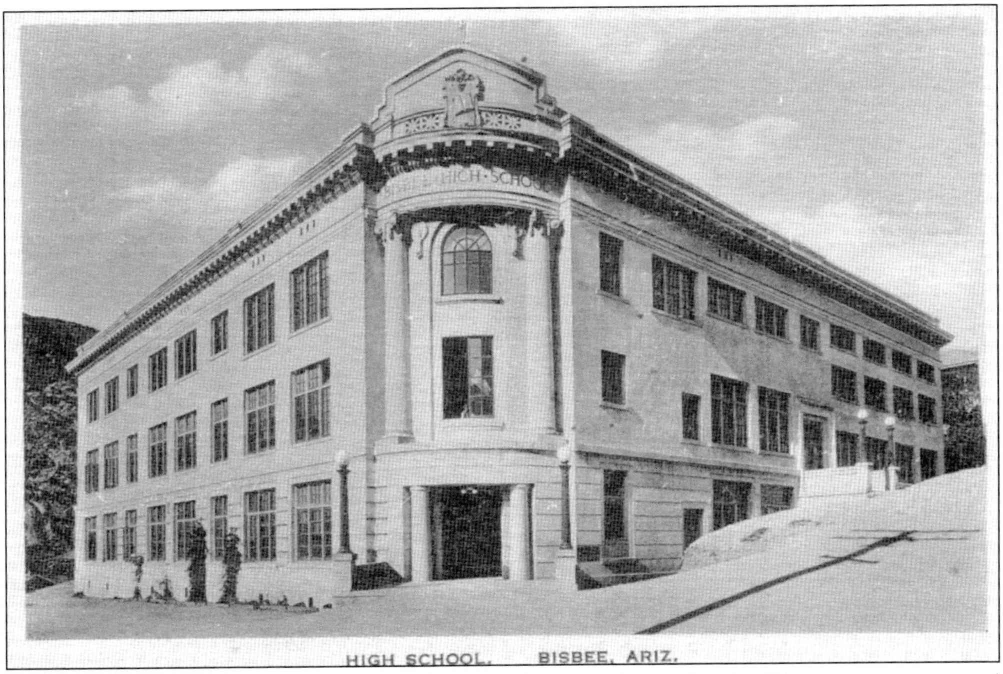

Central School housed Bisbee's high school students until another building was constructed in 1914 on nearby Clawson Hill. During the early morning hours of December 16, 1919, the school caught fire. The *Bisbee Daily Review* stated the blaze was caused by a defective furnace and left the building gutted. Despite the damage, the "foundation, walls and ceiling were not badly damaged" and were used when the school was rebuilt, allowing it to retain much of its original appearance, as seen below in the late 1930s. Bisbee's landscape offered an opportunity for the building to have an unusual feature. Despite having four stories, Bisbee High School has a ground entrance on each floor due to its being located on such a steep hill.

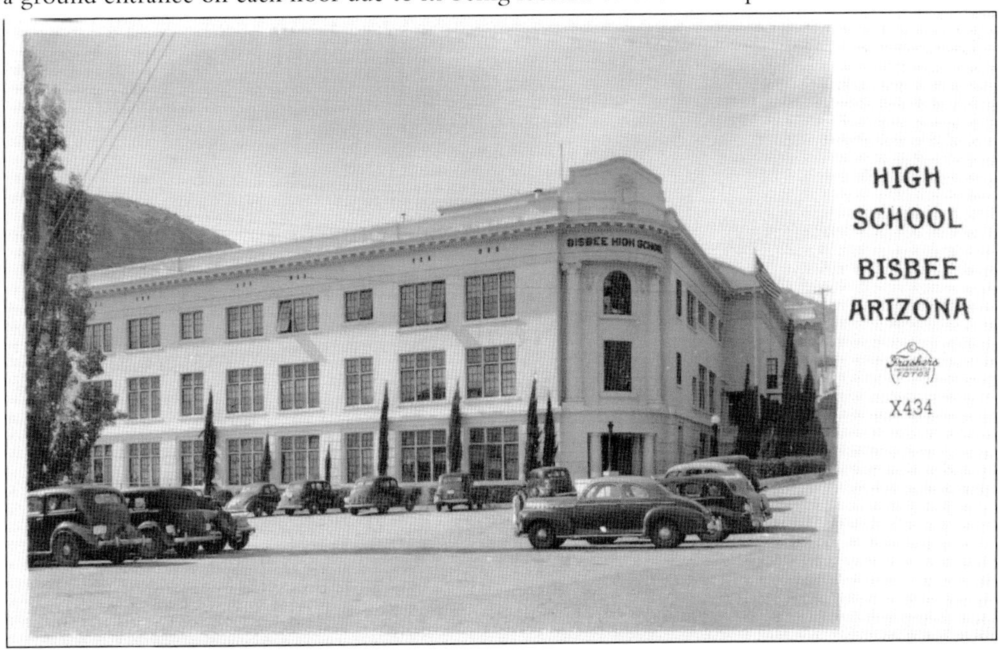

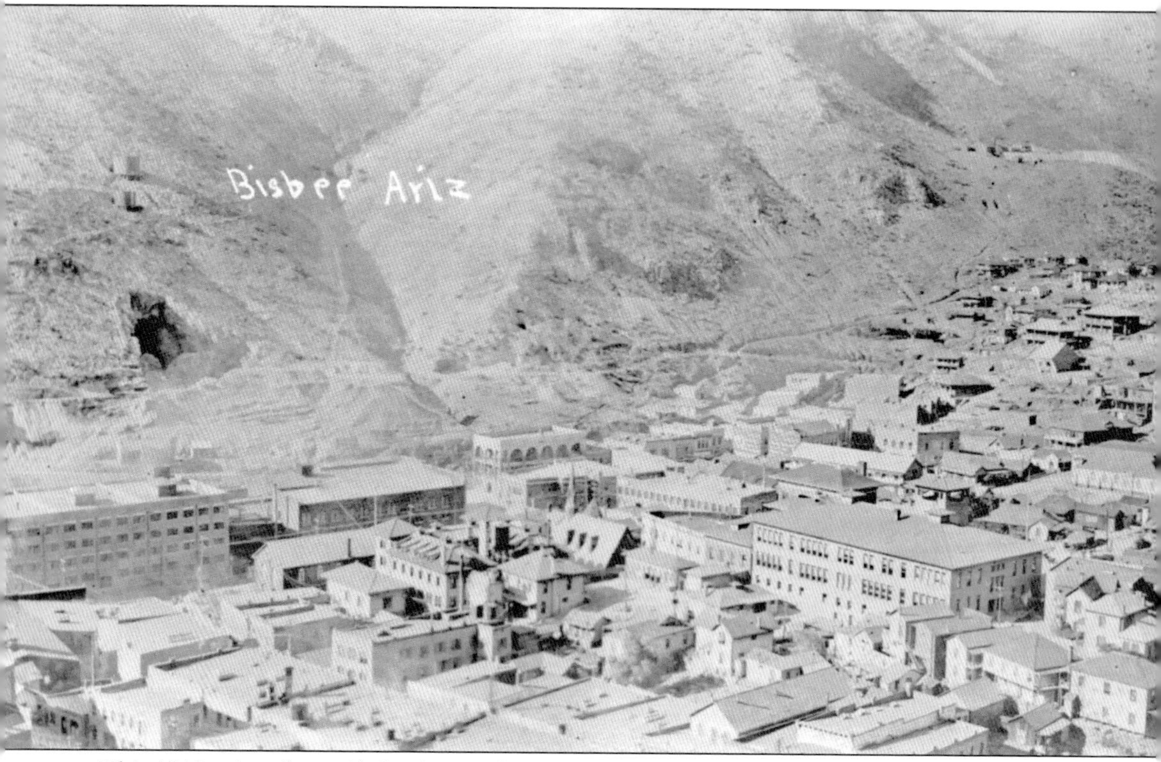

This 1910s view from Chihuahua Hill shows the Glory Hole on the left. The large opening in the mountain was where the first ore was removed from the Copper Queen Mine. The site yielded the richest ore body ever found in Bisbee, with an average grade of 14 percent copper. The Glory Hole also acted as a place of refuge for Bisbee's pioneers in the early 1880s. During this period, Apache raids were a legitimate threat to the mining camp. In the event the Apaches were seen in the surrounding hills, a mine whistle would blow four times to warn residents of potential danger. To seek shelter from a possible attack, Bisbee's women and children fled to the Glory Hole, which was always stocked with provisions, while the men of the mining camp situated themselves in the hills to protect the community. This drill was carried out multiple times, but fortunately for the residents, the Apaches never raided Bisbee.

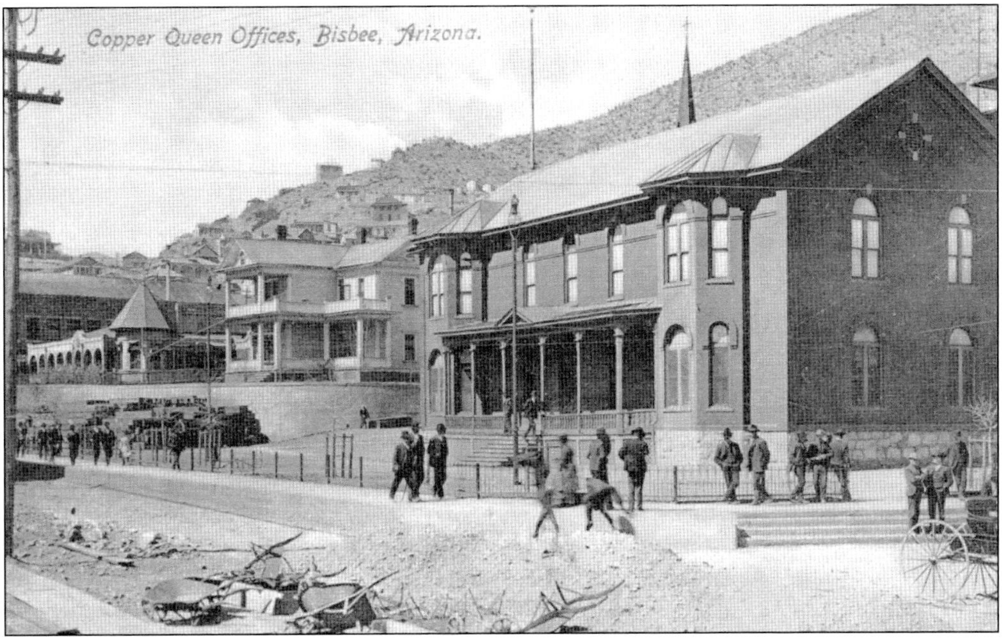

One of the buildings that had a view of the Glory Hole was the Copper Queen Consolidated Mining Company general office building. The structure was built in 1897 in the heart of Bisbee at the intersection of Brewery Gulch and Copper Queen Plaza. Today, the building houses the Bisbee Mining & Historical Museum.

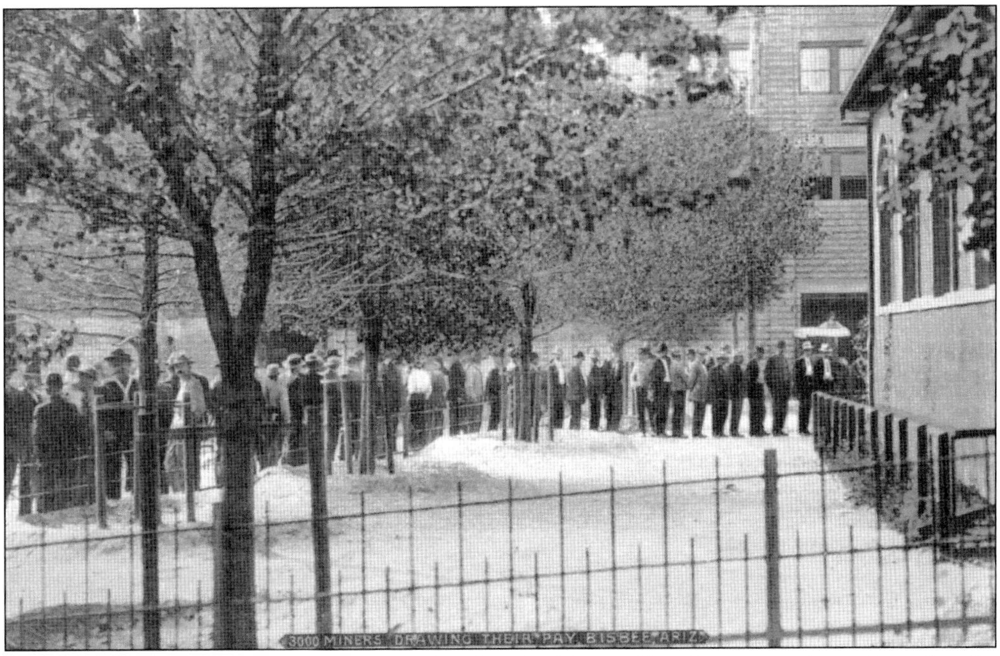

The centralized location of the general office building offered a convenient place for the company's paymaster office, and an addition was built on the east side of the building to house the department. Each payday, employees of the mining company lined up to receive their earnings, a practice that occurred for decades and was witnessed by countless residents.

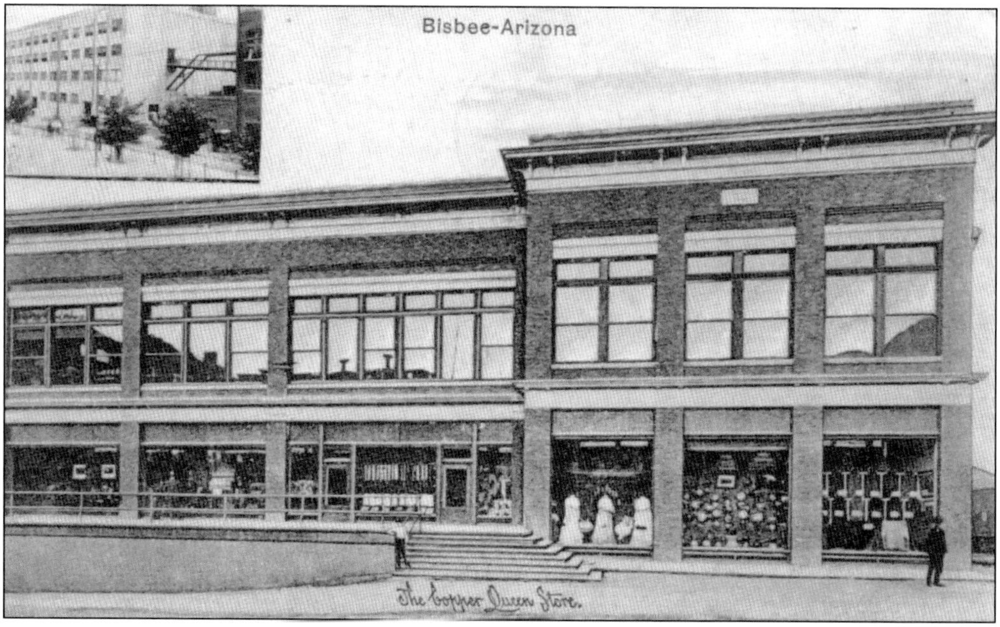

In an effort to retain a reliable workforce, the mining companies brought in the amenities found in any modern community to meet the needs of residents. This included providing a company store, shown above in the 1910s, where employees could purchase some of the finest wares available in the area. Due to the severe topography of Bisbee, many of the stores provided a delivery service for their customers. Homes sat high up on the hills and often up over 100 stairs. Rather simple chores, such as a household's shopping, could quickly turn into a nightmare for a lady attempting to take her purchases home. These delivery wagons, seen below in 1902, enabled customers to purchase what they needed without having to concern themselves with how they would transport the items home.

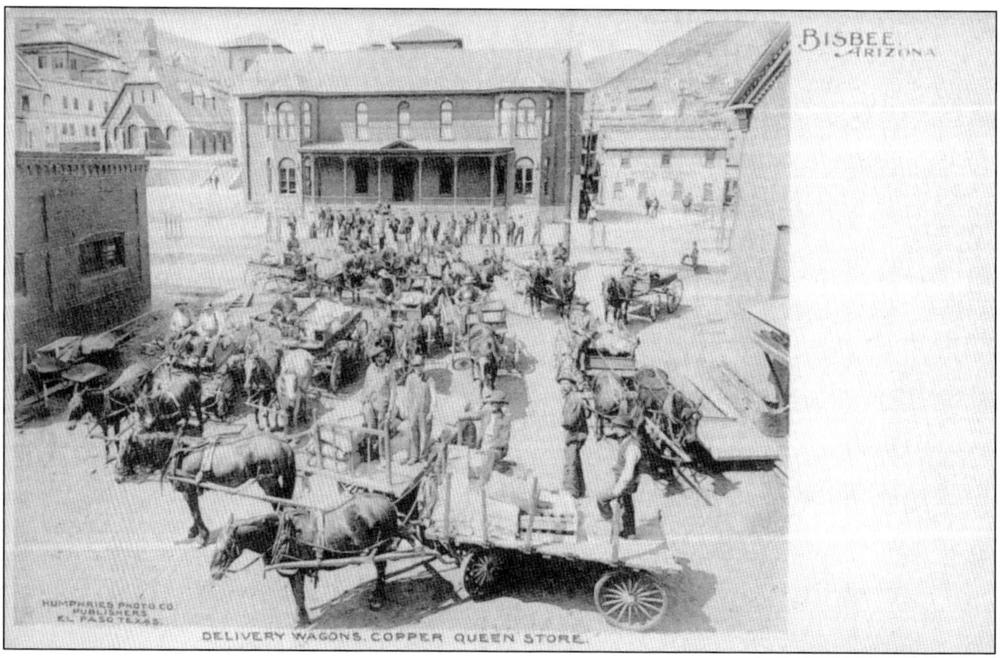

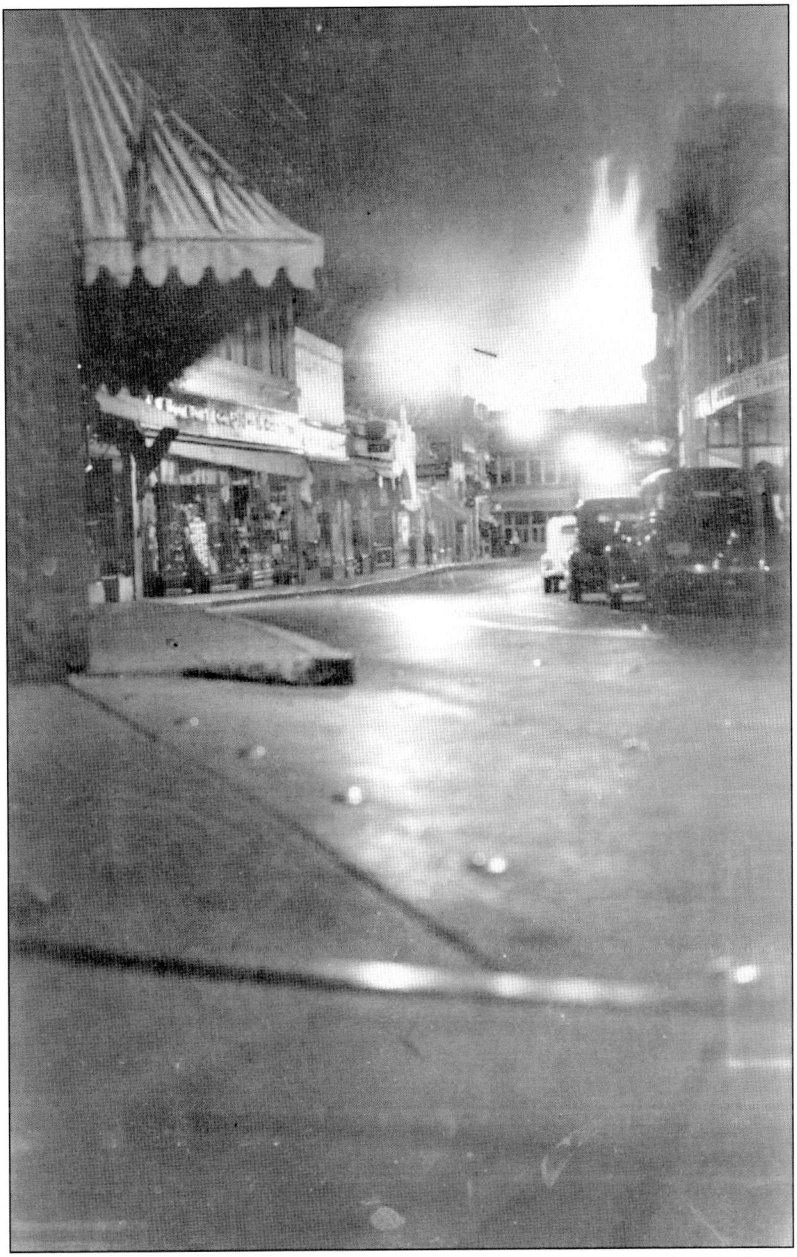

Just after midnight on September 23, 1938, a fire swept through the Phelps Dodge Mercantile Building. The *Bisbee Daily Review* reported that fire trucks from both Bisbee and nearby Lowell responded to the call, but the fire was too great for the firemen, as it became a "roaring inferno." Within an hour of the blaze, residents flocked to the site to witness the event, many of whom were reported to have left their homes so quickly they were wearing their pajamas. As the intensity of the fire increased, the scene "resembled a battlefield as walls began to tumble and crash with a roar, huge show windows shattered and splintered, bottles in the grocery department broke with a loud pop and ammunition in the hardware department was set off resembling machine gun fire in an armed conflict." The increasing danger of the situation caused onlookers to be moved far back by officers as the firemen worked to contain the fire.

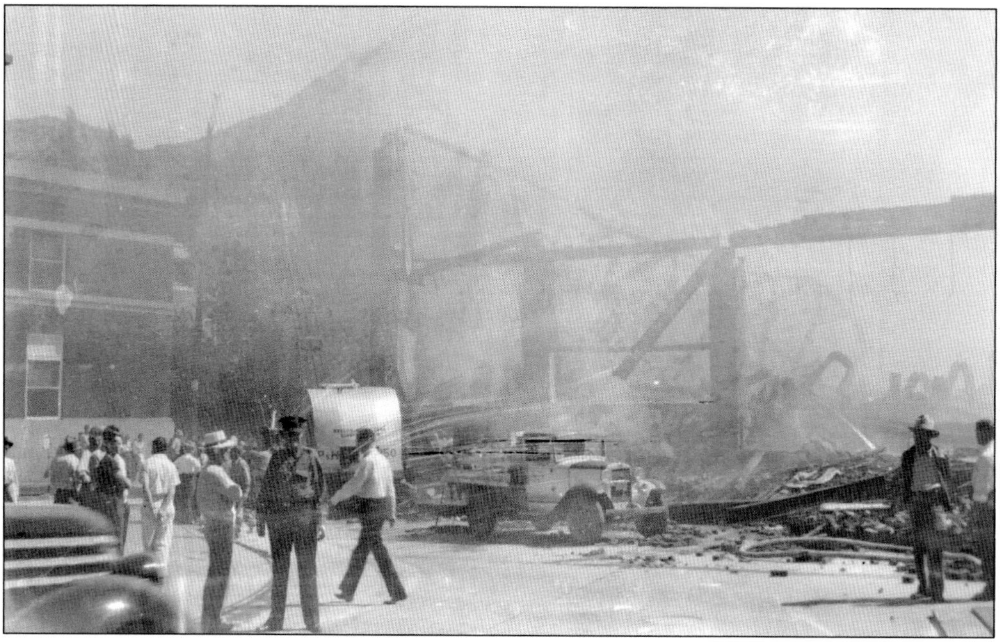

By the time the sun rose, only ash, rubble, and twisted metal beams remained where the structure once stood. Employees of the store, now without jobs, were said to be standing by and staring at the smoldering ruins of the building. At the time, the business was one of the largest department stores in the Southwest and sustained a loss estimated at over $300,000. Being an important fixture in the community, Phelps Dodge rebuilt the mercantile, which reopened for business in December 1939. The undertaking was completed by famed housing developer Del Webb and was his first project.

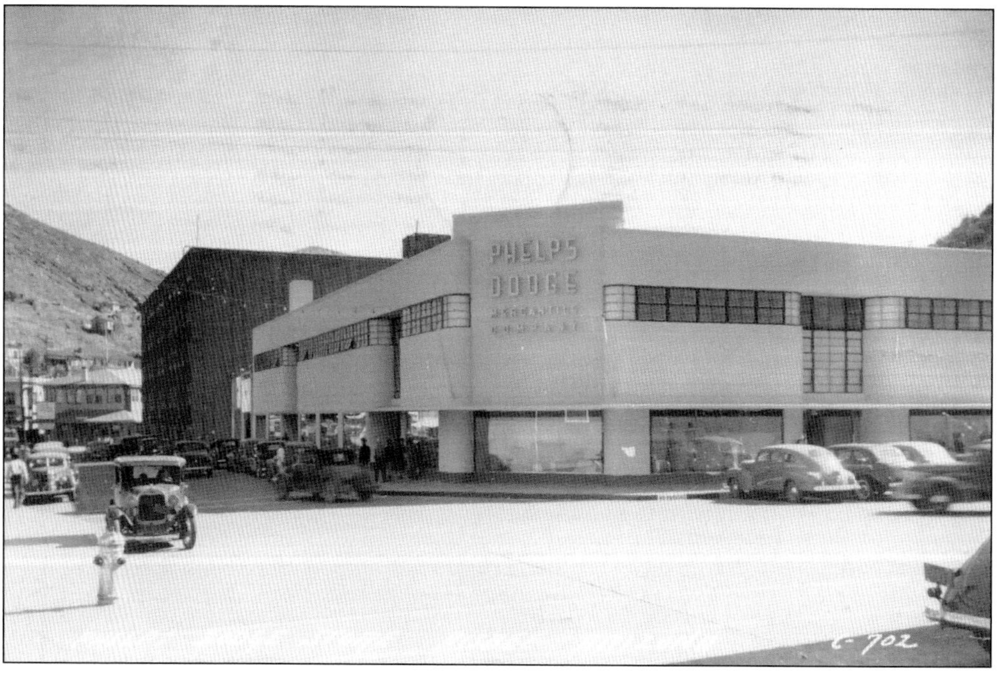

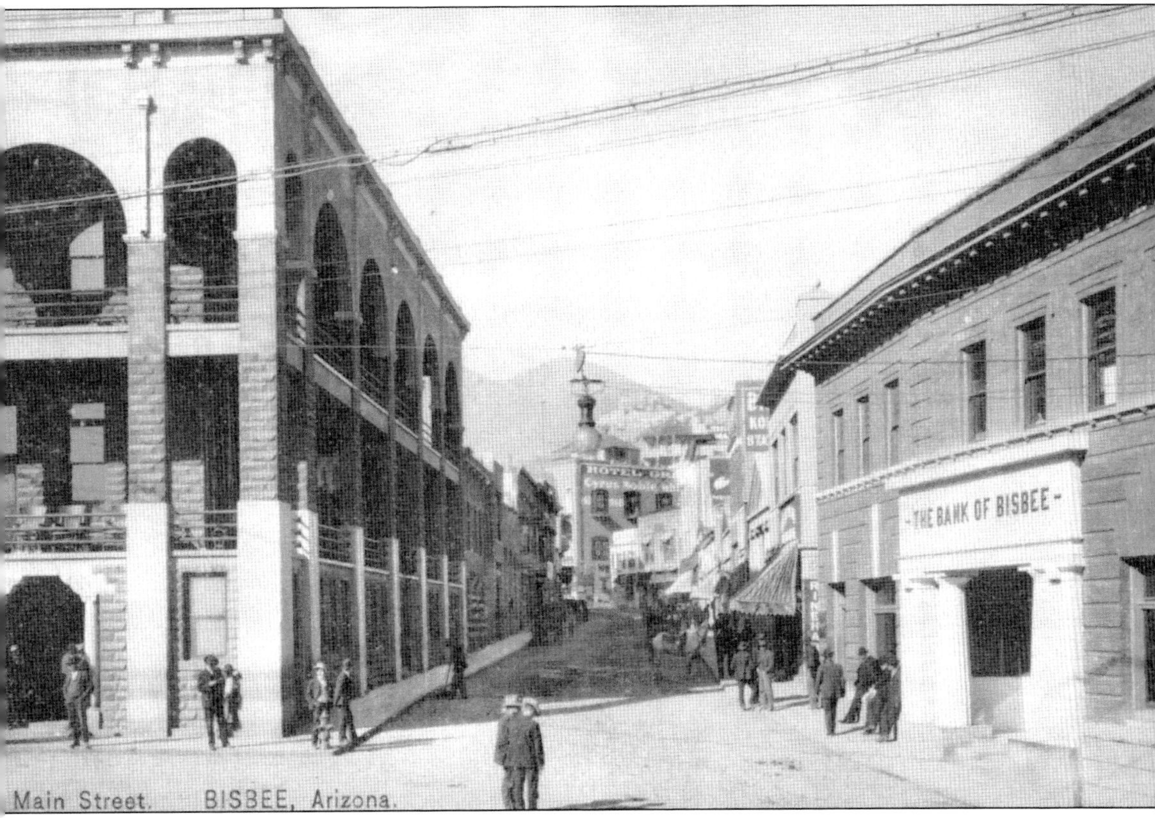

Bisbee's Main Street begins just across the street from the mercantile building. This 1907 postcard illustrates the great progress the community made just 30 years after the first discovery of ore in the Mule Mountains. The seemingly inexhaustible ore deposits found in Bisbee lured miners to the Mule Mountains, with merchants quickly following to provide supplies for the men and their families. By this time, the city also had a well-established library, shown on the left, and a bank, on the right. Lodging facilities, barbershops, clothing stores, and drugstores were also located on Main Street. However, saloons were the predominant business located along the mining camp's thoroughfare. The 1906 Sanborn Map for the city reveals no less than 12 drinking establishments located on this section of Main Street alone.

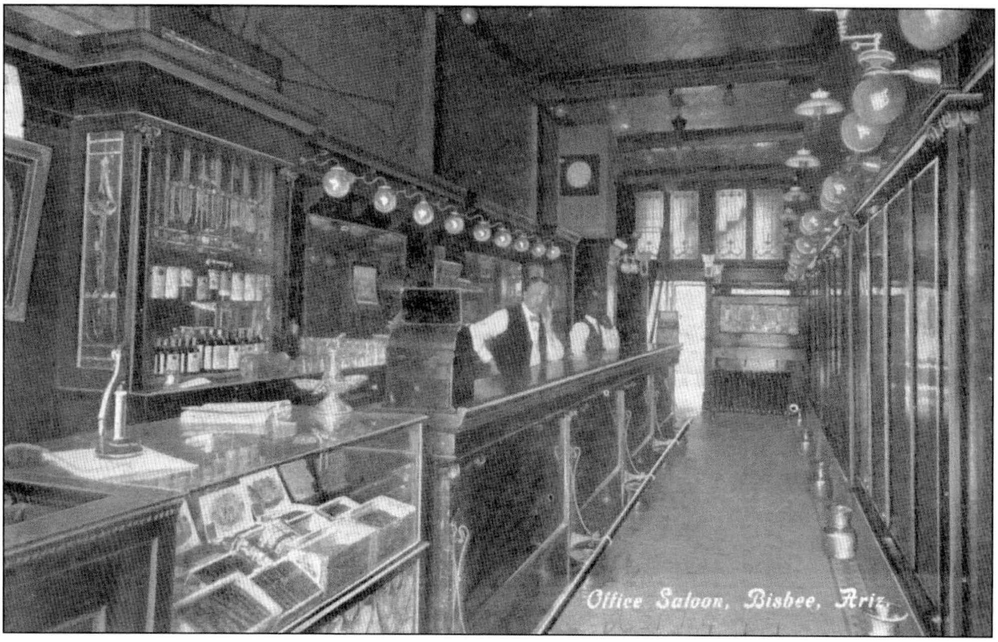

The Office Saloon, shown above around 1910, was one of the locations on Main Street where a miner could stop in to have a drink and smoke a cigar. The back of the establishment's postcard advertised they also served a "Classy Merchants [sic] Lunch" daily in their club rooms on the second floor. In addition to finding sustenance for the body and spirit, residents could find a wide array of goods to purchase on Main Street. One such location was the Bisbee Drug Company, which offered the latest remedies to cure a person's ills, as well as soap, bibles, and "perfumery of the best sort."

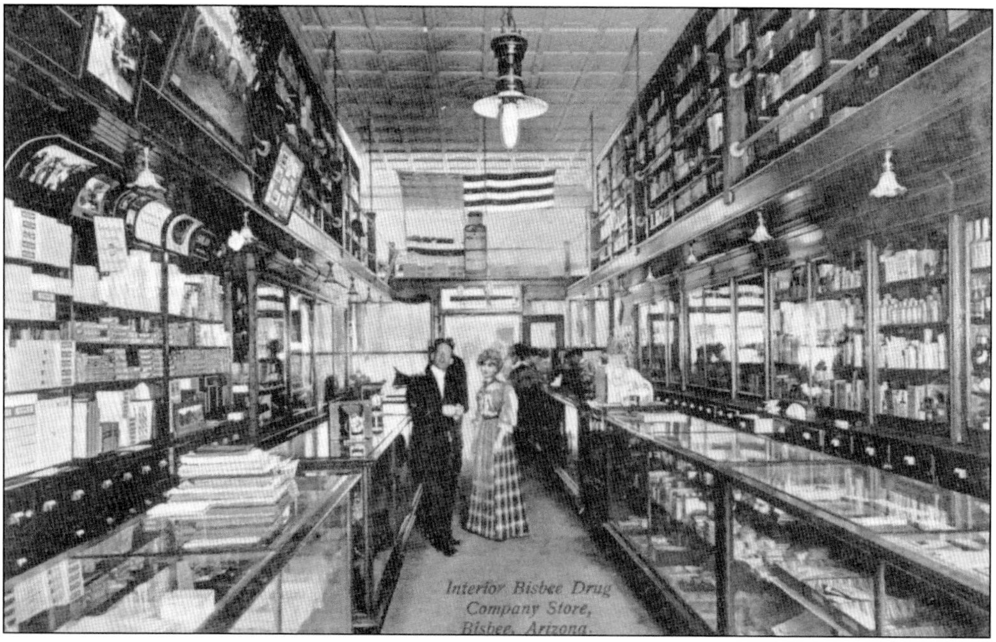

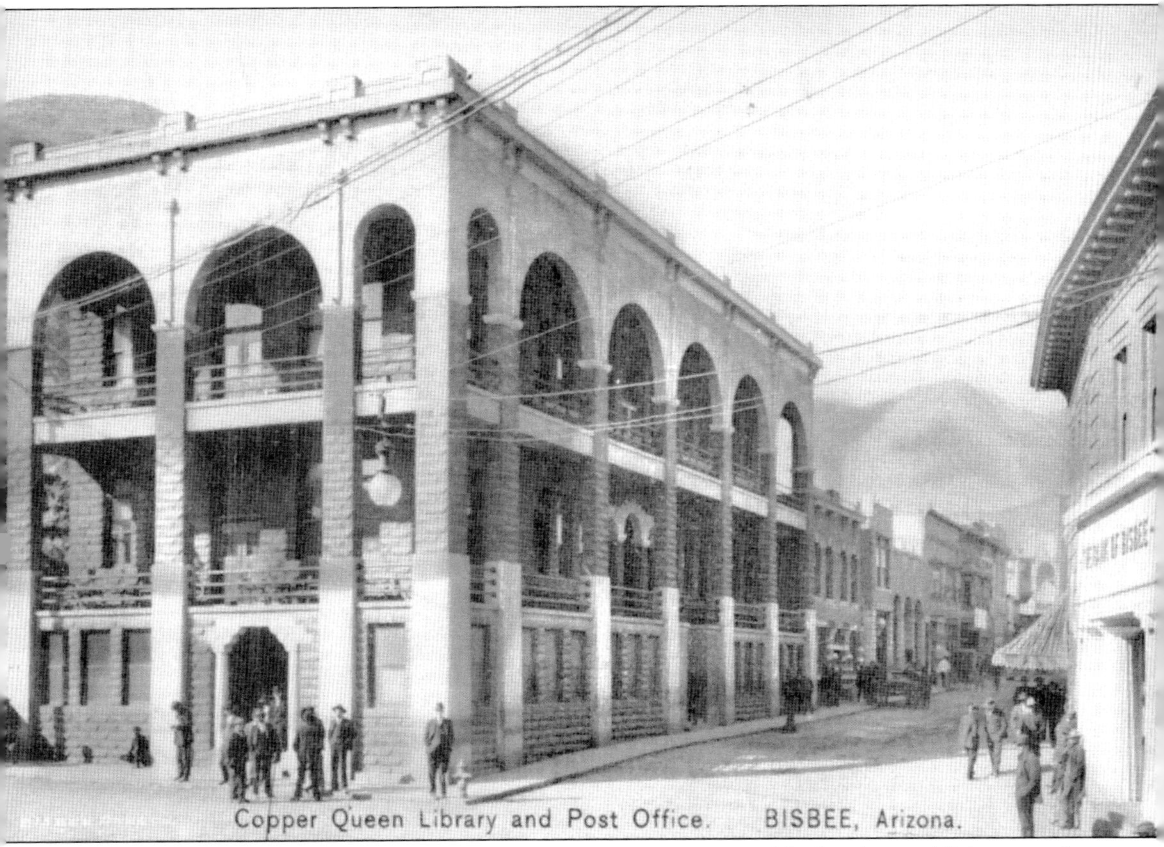

Copper Queen Library and Post Office. BISBEE, Arizona.

Local legend states that an 1884 lynching in Bisbee was responsible for the establishment of the mining camp's Copper Queen Library. Upon entering Bisbee, Ben Williams, an executive of the Copper Queen Mining Company, witnessed a lifeless body hanging from a tree limb near Castle Rock. Believing the only way to combat such uncivilized behavior was through educating the public, he sent a large selection of books to Bisbee that were available to be checked out by residents. The service became so popular that the mining company erected a library in the city in 1887. With time, the book collection grew, and so did the facilities. By 1907, the company built a third library building large enough to accommodate the ever-growing collection. The library was housed on the second and third floors of the new structure, with the first floor being home to the local post office. Recognizing that Bisbee's mountainous topography would break even the most determined postal employee, residents were required to obtain their mail directly at the post office, a practice still in place today.

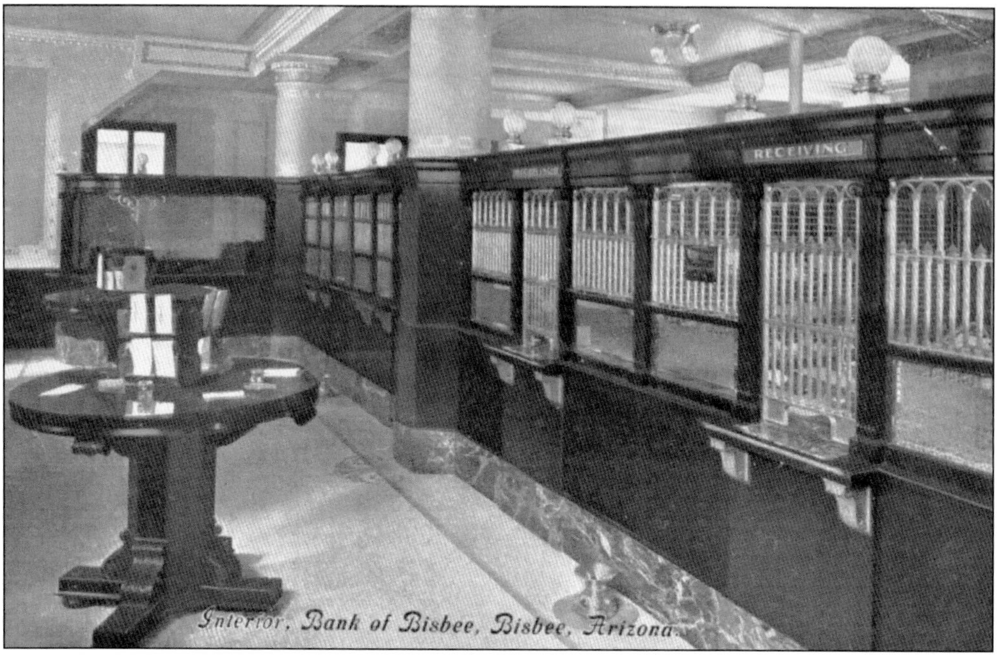

Interior, Bank of Bisbee, Bisbee, Arizona.

An increase in population and commerce warranted a need for a sound banking institution in Bisbee. During 1900, the Bank of Bisbee was formed by prominent members of the community and was an important economic indicator for the city. In 1902, the *Bisbee Daily Review* stated, "There is nothing that speaks for the prosperity and general flourishing condition of a community as a well regarded banking system." By this time, the Bank of Bisbee boasted it held $750,000 in deposits and erected a "beautiful" brick building at the beginning of Main Street, reflecting its stature in the community. The greatest source of pride for bank officials was the vault. The bank's president stated, "When money is deposited in that vault at night, we propose to find it there in the morning. It is absolutely burglar proof."

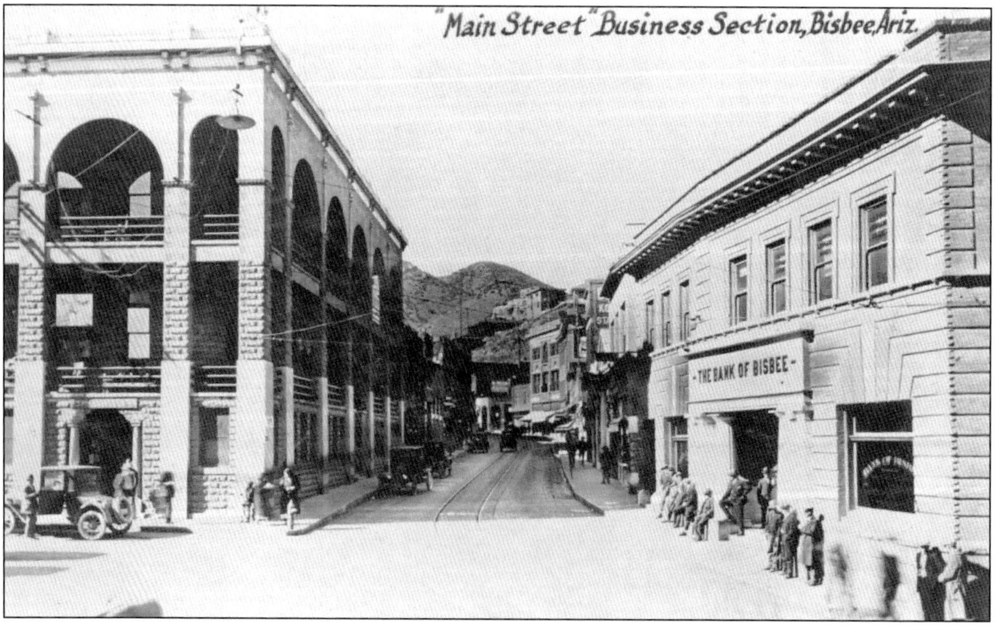

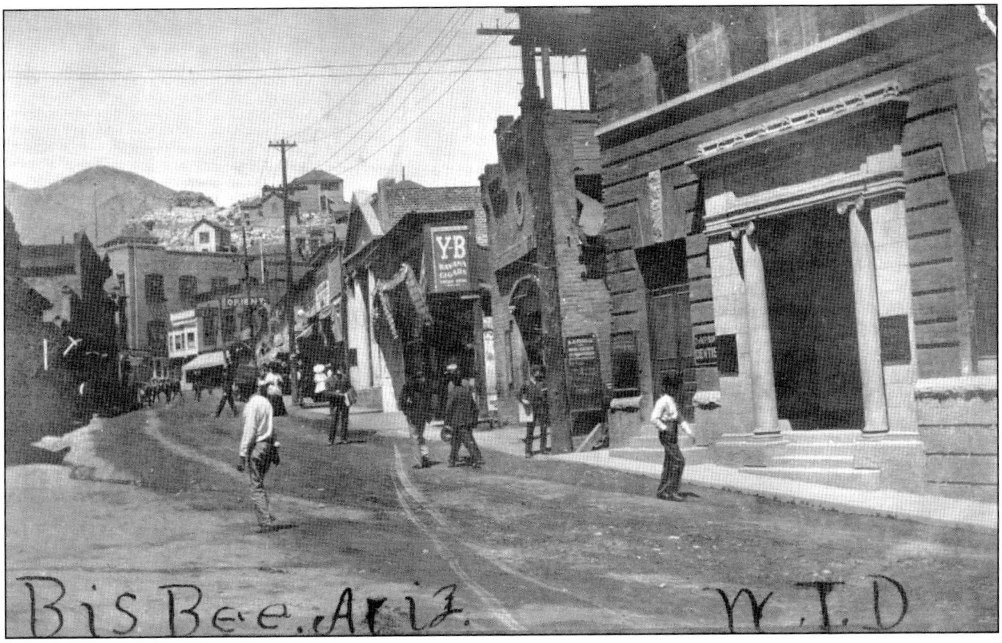

With time, the face of Main Street began to change. This transformation is easily seen in the city's postcards, such as these mailed in 1905 (above) and 1908 (below). Wooden structures were removed to make way for larger, more permanent brick buildings. In 1907, the *Bisbee Daily Review* noted, "One by one the old landmarks of the city are disappearing before the march of progress, and in their stead are erected structures in which are embodied, inventions, materials and workmanship which are the product of more advanced ideas. To become aware of this fact, one has but to take a look at Main Street, where on several new buildings have already been erected, and today preparations are being made for the construction of several others."

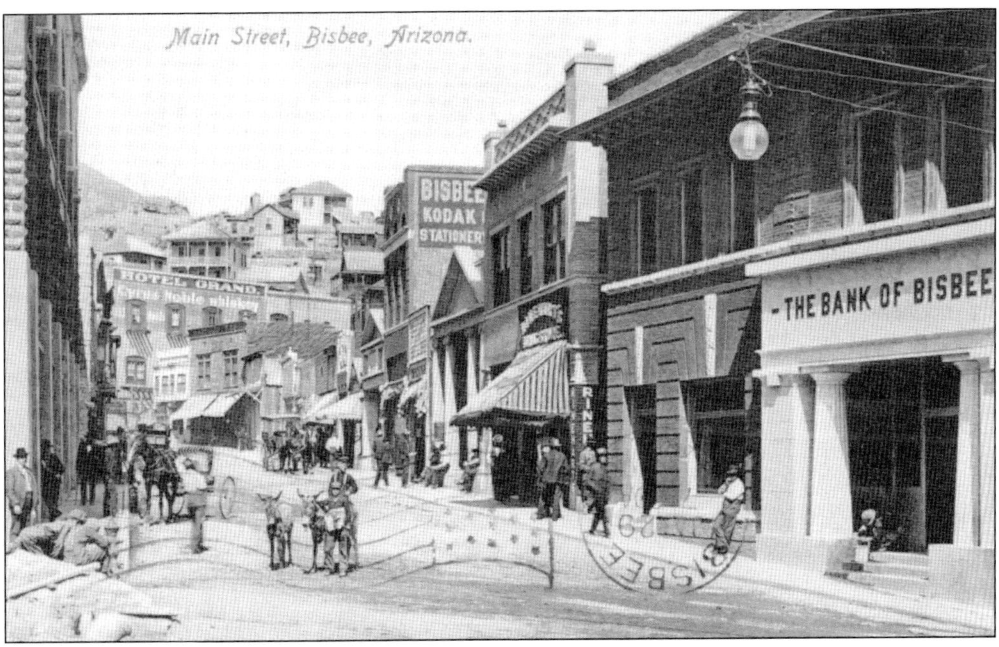

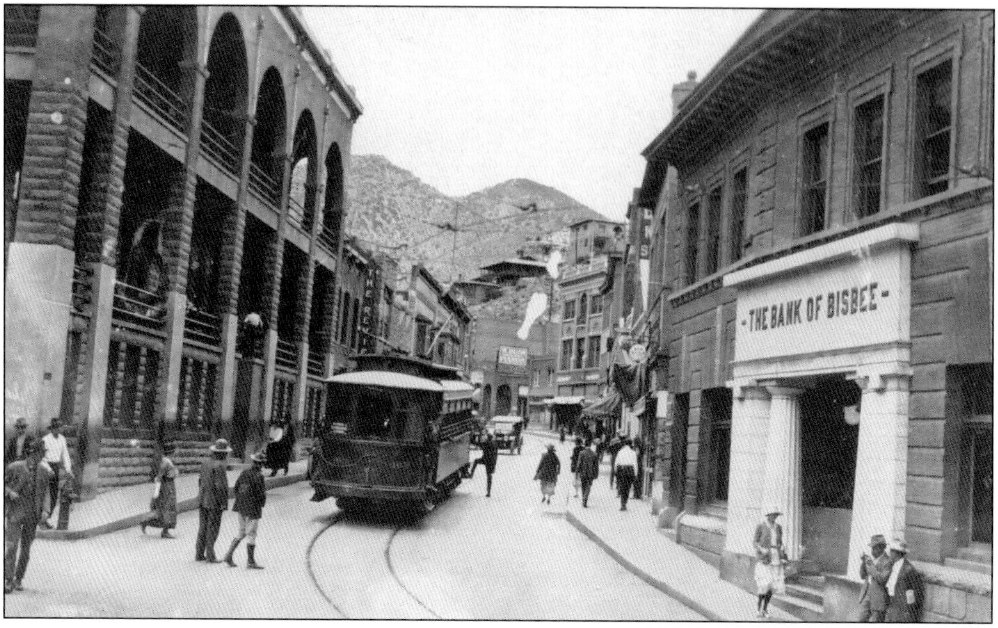

Bisbee's muddy Main Street vanished, and a paved thoroughfare was put in place, as shown in these cards from the 1920s (above) and 1930s (below). The street that was once a path for horses and wagons became a road for automobiles and a streetcar. The wooden buildings on the streets disappeared and were replaced with structures made from lasting materials such as brick. Much of Main Street was also rebuilt after the devastating fire of October 14, 1908. Flames consumed half of the establishments that lined the road, with the upper portion of Main Street left in ruins. However, most business owners were quick to rebuild, and from the ashes rose structures described by the *Bisbee Daily Review* as "some of the best buildings which had been erected in the city."

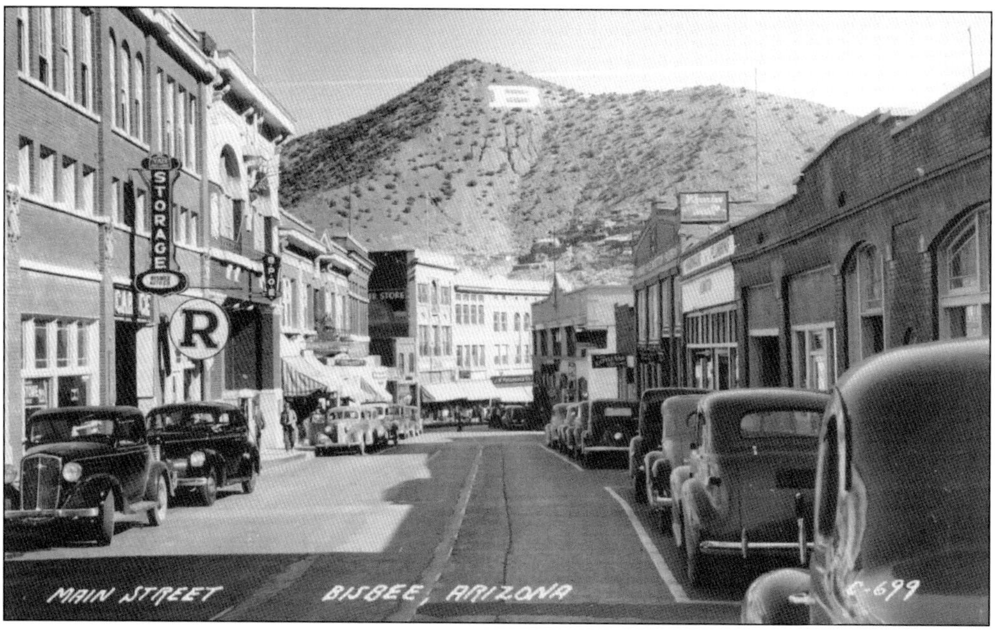

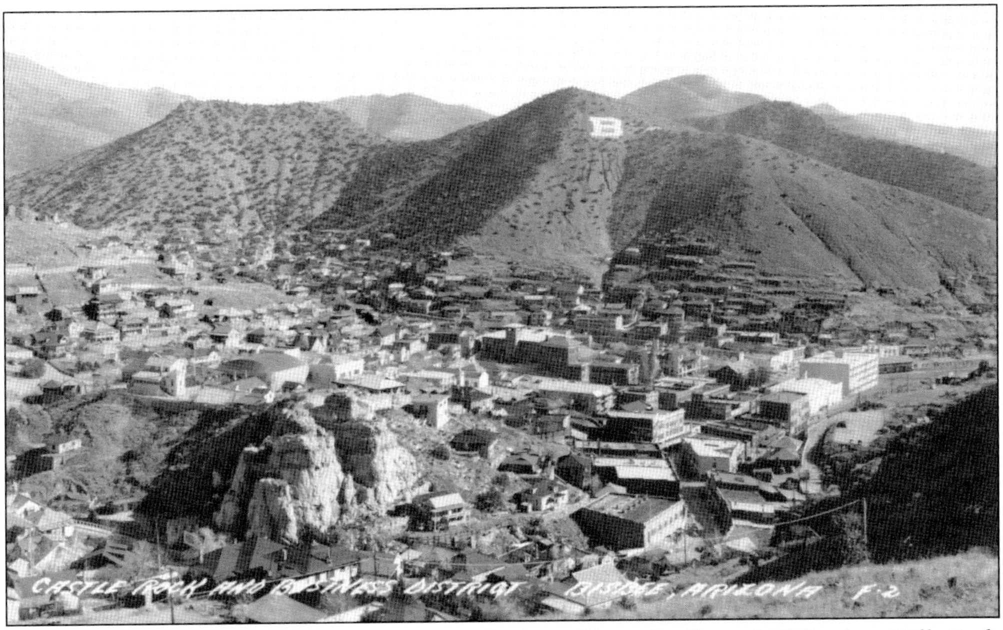

In 1928, Bisbee High School students created a large B on the face of Chihuahua Hill. Each year, incoming freshman were charged with maintaining the appearance of the B, a symbol of school spirit and community pride, by adding fresh lime and paint to its surface. The B was also lit on fire the night before the high school's annual football game with the rival smelter community of Douglas.

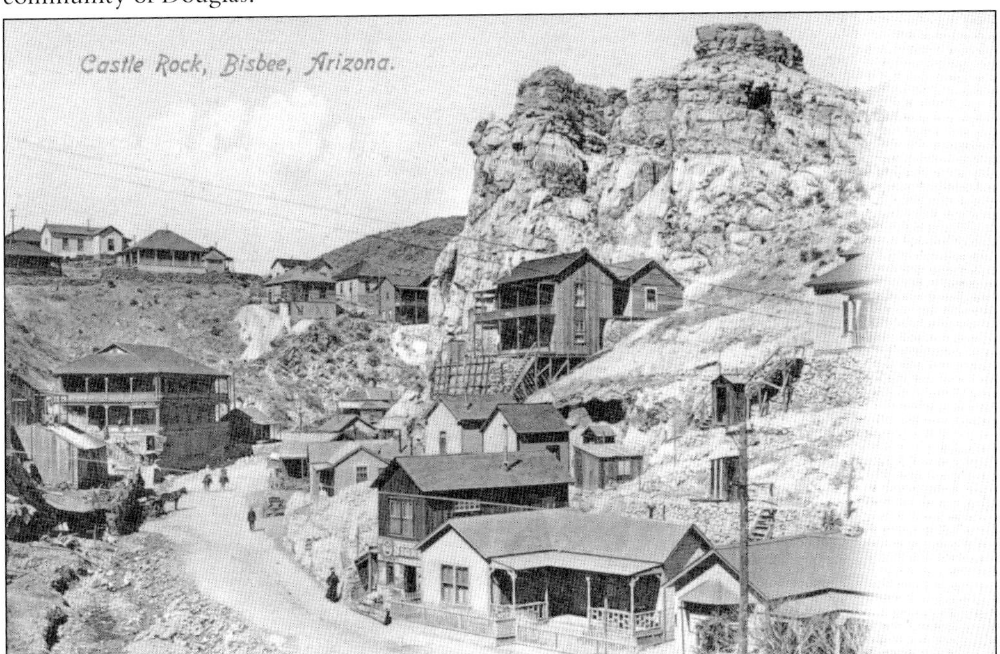

As Main Street ends, Tombstone Canyon begins. This transition is noted by Bisbee's best-known geological landmark Castle Rock, shown here around 1905. Historic maps mark the limestone mass as towering 120 feet above the base of the canyon floor.

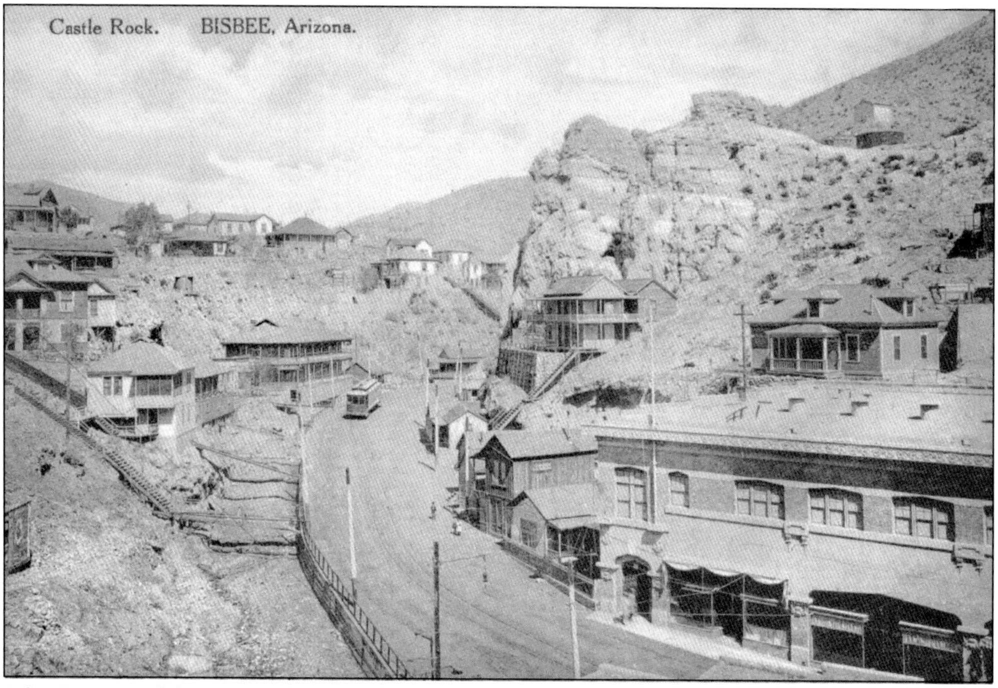

John Dunn and the US Cavalry scouting party, who first discovered ore in the Mule Mountains in 1877, camped at a spring across from Castle Rock. The spring was located at the site where the Muirhead house was erected, which is to the left of the trolley in this 1910s postcard.

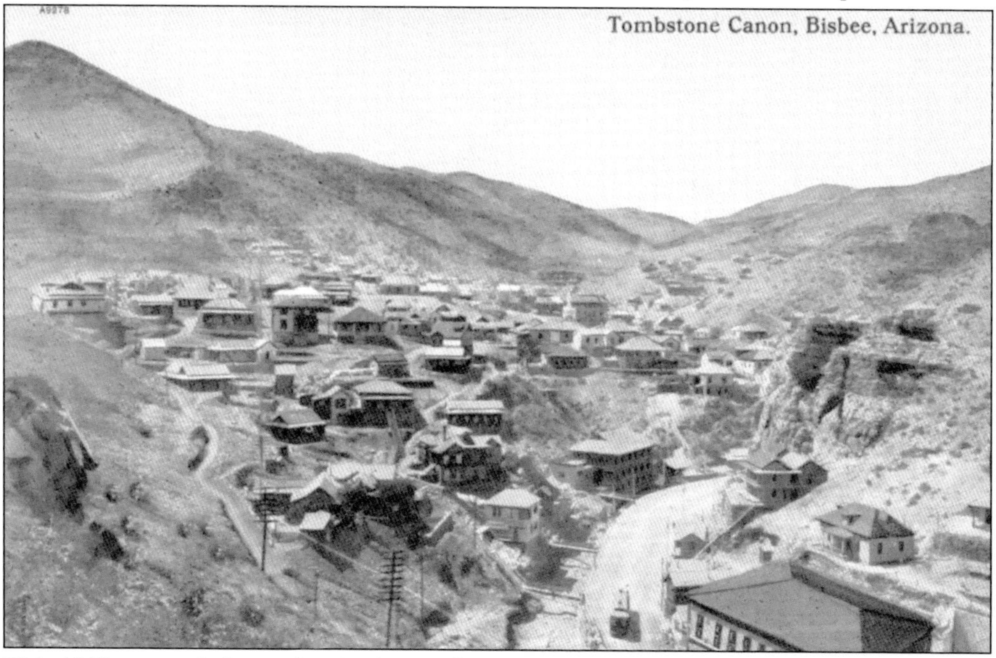

This 1910 postcard shows the Quality Hill neighborhood located across from Castle Rock. The area was chosen as the site to build houses for senior management of the mining company. However, like other people who lived in Bisbee, many residents of Quality Hill still had to climb stairs to reach their homes.

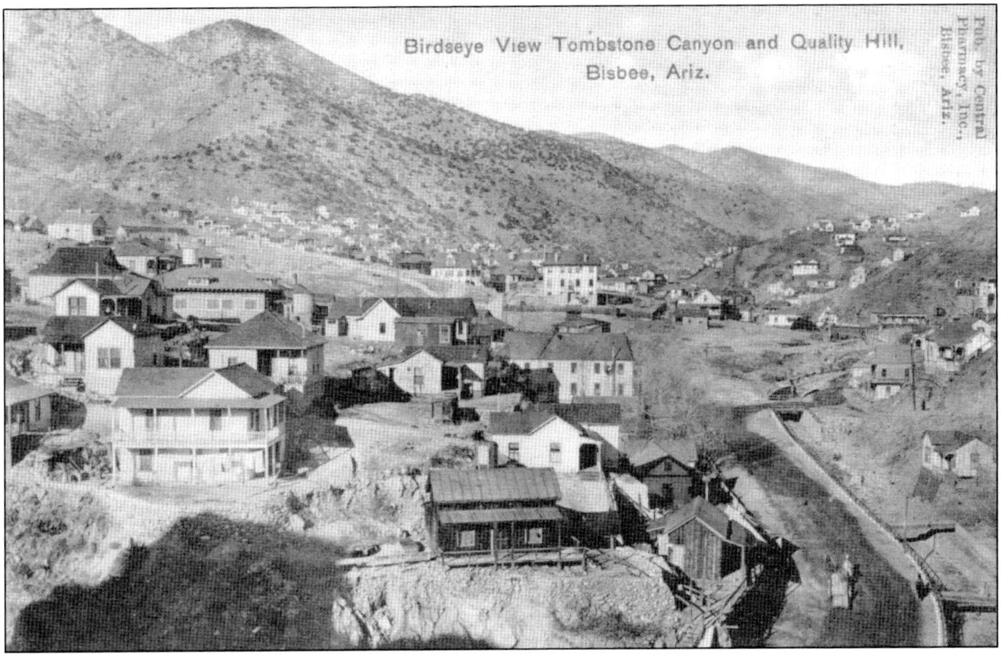

This 1910s postcard illustrates how Tombstone Canyon continued up along the base of the mountains to the west. The canyon received its name by being the route frequently used to travel to the nearby mining camp of Tombstone.

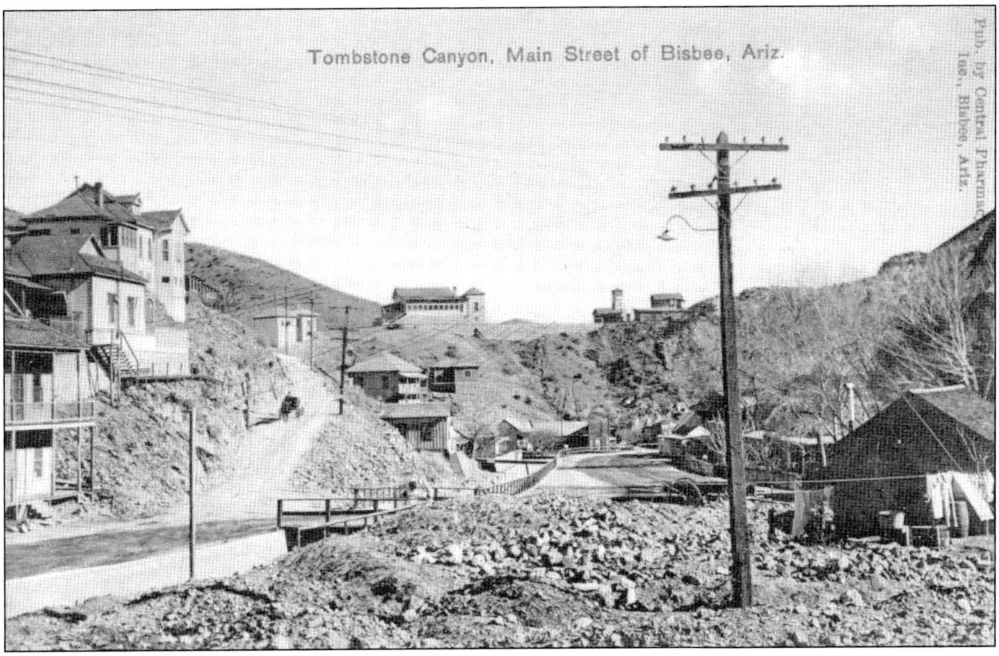

Flooding was a serious problem for the inhabitants of Bisbee. Torrential monsoon rainwater quickly flooded the canyons, carrying with it dirt, rock, and debris. In an effort to combat this problem, a ditch system, seen on the left of this early 1900s postcard, was created to channel the water off the streets and away from homes.

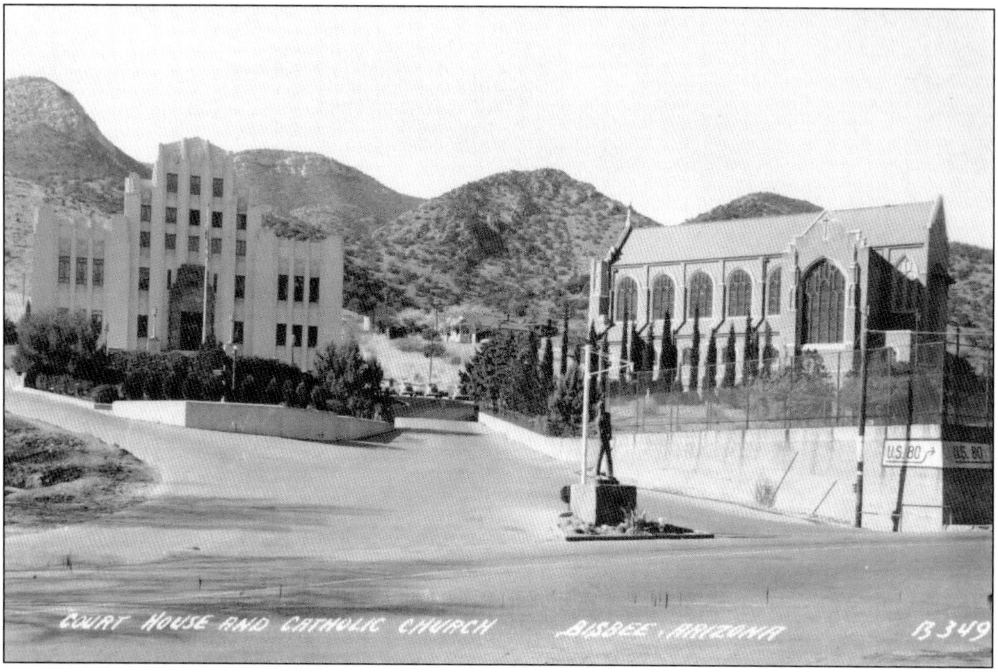

Continuing up Tombstone Canyon, one will find the Copper Miner statue, pictured here about 1940. The statue was part of President Roosevelt's New Deal program and was funded by the Federal Emergency Relief Administration in 1935. With an imposing stance and hands firmly gripping a four-pound hammer, the statue honors "those virile men, the copper miners," of Bisbee. Not only is the statue worthy of note, but also the block of Gunnison granite it rests upon that was used in the legendary hand-drilling contest on July 4, 1903. During the contest, a team used a set of hand drillsteels and an eight-pound hammer to drill 46 11/32 inches into the hard rock in 15 minutes, a record that stands to this day.

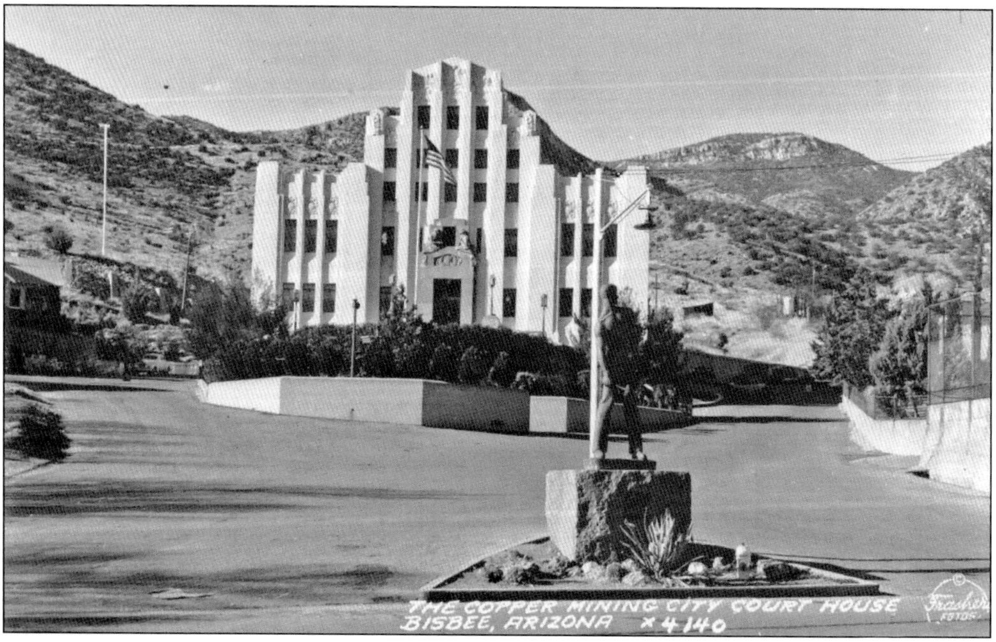

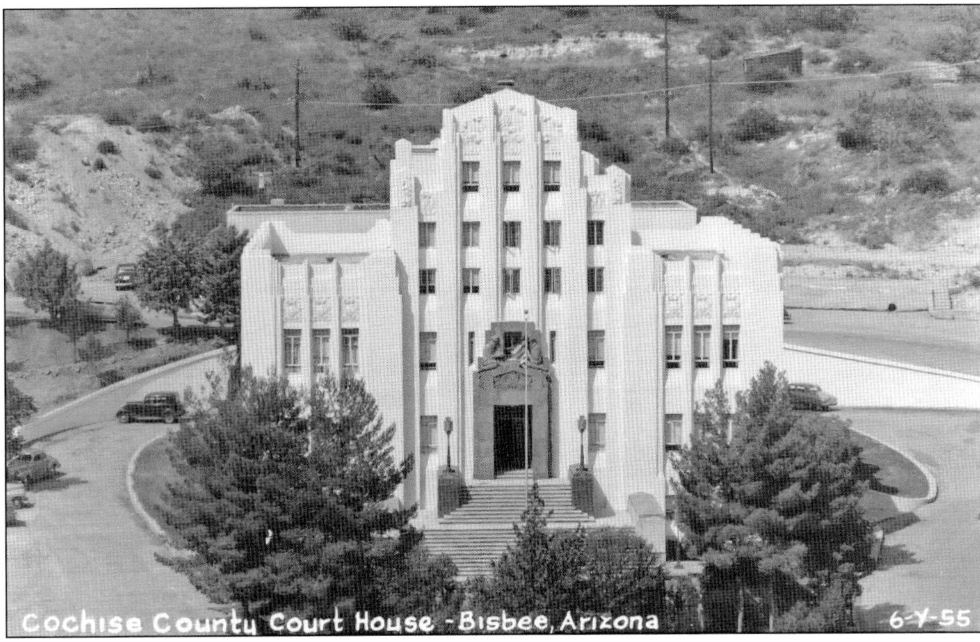

The Cochise County Courthouse is located directly behind the Copper Miner statue. During 1929, voters approved to move the seat of county government from the dwindling silver mining camp of Tombstone to Bisbee, where the mines continued to thrive. Upon opening in 1931, the *Bisbee Daily Review* stated the courthouse was a "beacon of confidence to the future of Arizona's greatest and most stable copper camp." The Art Deco structure, shown here around 1940, was comprised primarily of "native Arizona materials and [contained] more copper work than any other public building in the western states." The exquisite doors are the best-known copper feature of the building; the doors are crowned with a stone carving depicting two prospectors panning ore.

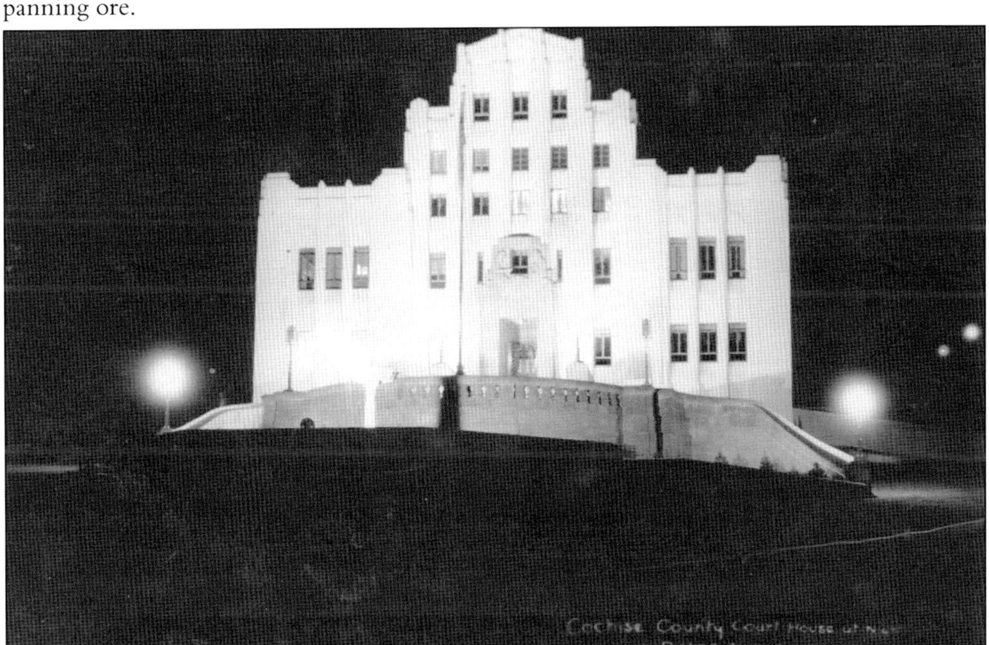

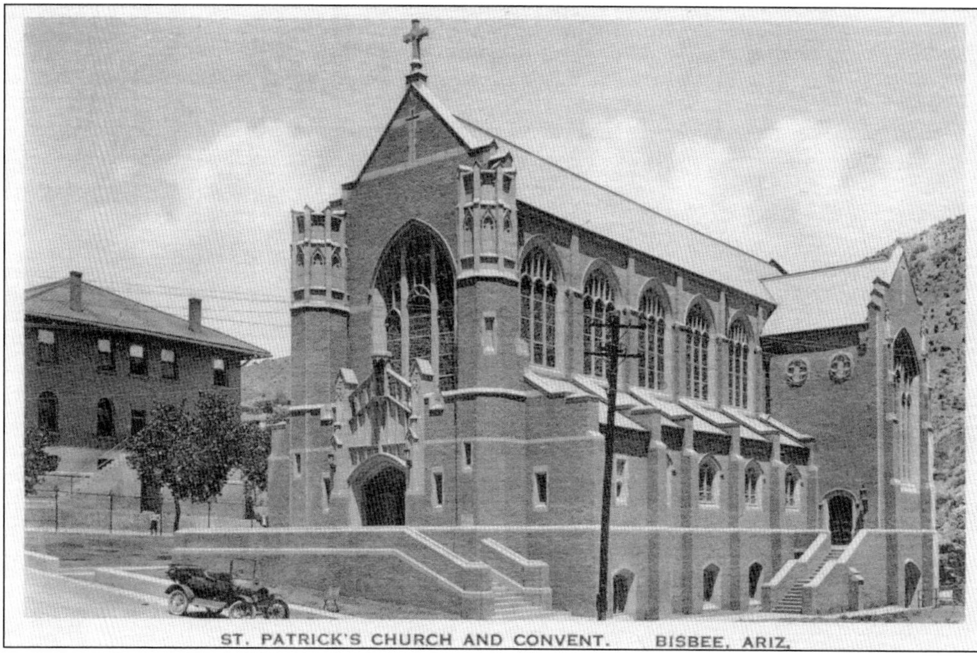

The 1930s postcard above captures Saint Patrick's Catholic Church. The land for this house of worship was donated by Thomas Higgins, who became successful through area mining claim transactions. The church contains 11 exquisite stained-glass windows that illustrate the Assumption and events in the life of Jesus. Next to Saint Patrick's Church is Loretto School, shown below. The Catholic school opened in 1907 and was taught by the Sisters of Loretto. In 1917, around the time this postcard was printed, the Bisbee *Daily Review* noted that the "Discipline of the school and its refined environment are widely known throughout the southwest."

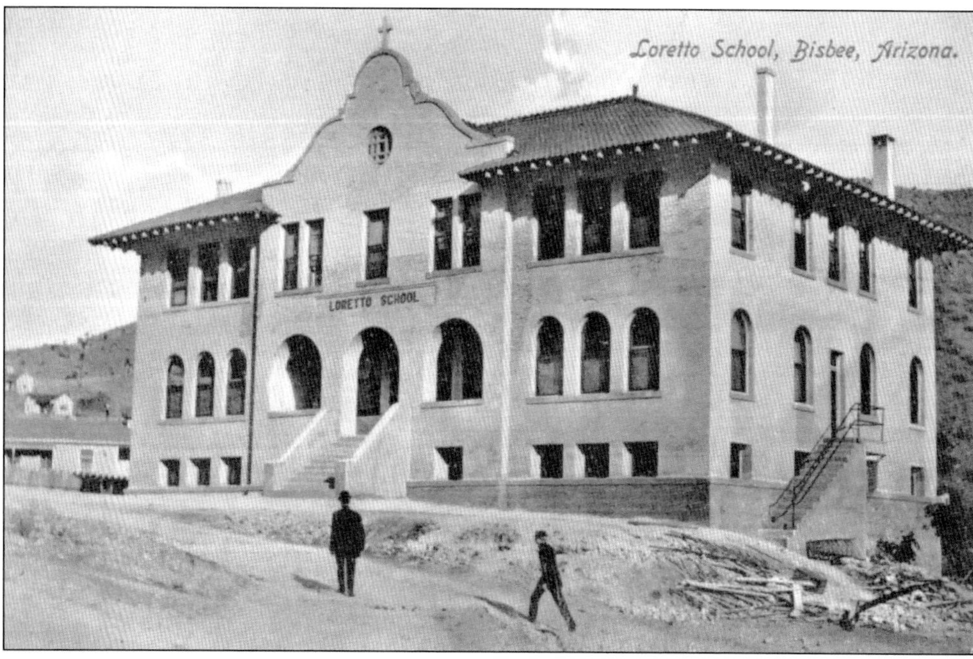

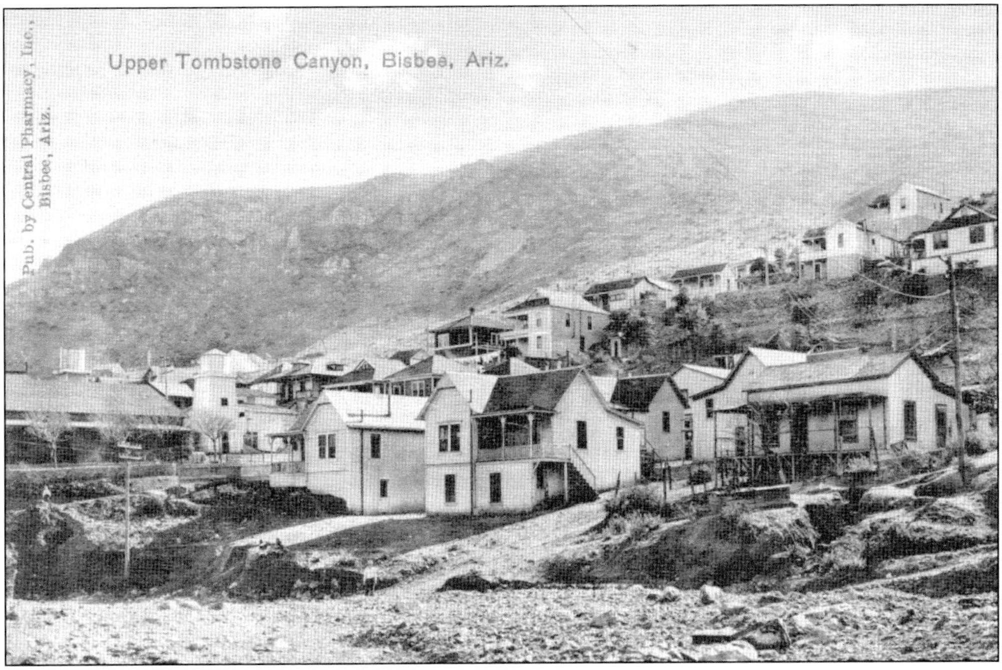

One of the indicators of Bisbee's prosperity was the growth of the residential districts, as illustrated in these 1910s images. As more people moved into Bisbee, homes were built up along the floor, as well as on the hills that sprawled up from Tombstone Canyon. The inhabitants of these homes, often families, were another sign of the community's success. Families represented a stable workforce, as opposed to the transient nature of many single miners. Because the number of families who lived up Tombstone Canyon grew, a fire station, schools, and grocery stores were erected in the area to fulfill the needs of these residents.

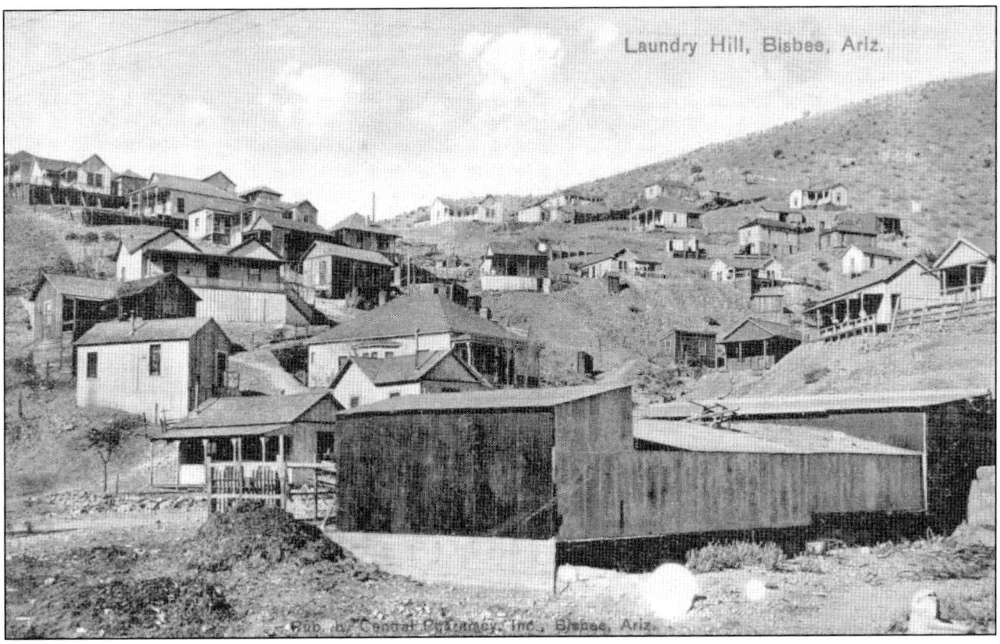

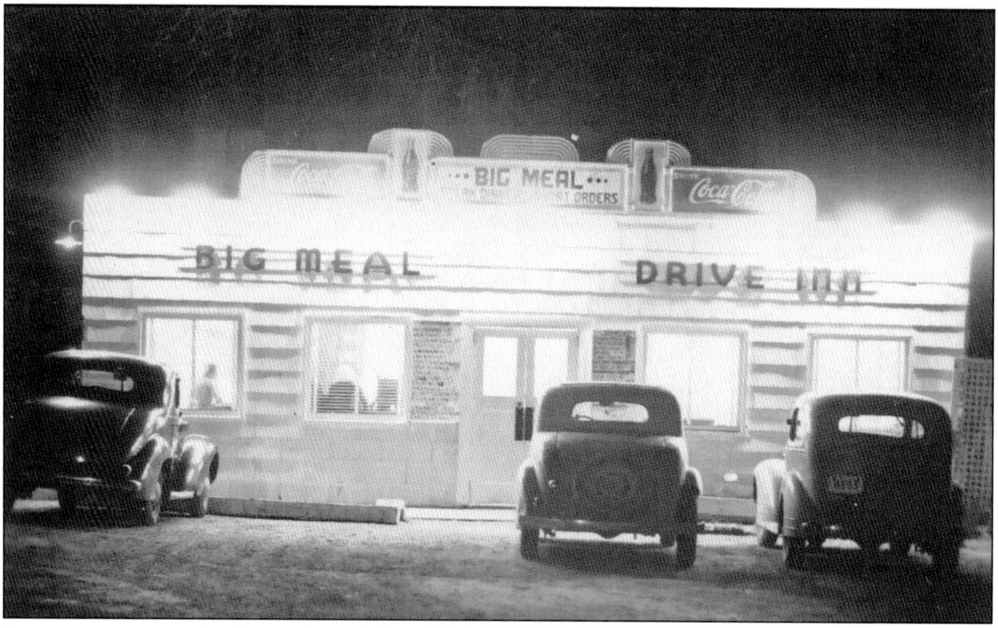

Restaurants were also located up Tombstone Canyon, including the Big Meal Drive Inn, later known as the Arctic Circle. The restaurant was a local favorite known for its delicious five-for-$1 hamburgers. The Big Meal Drive Inn, pictured here in the 1940s, was also a signal of changing times, as more people acquired personal automobiles for travel and became less dependent on public transportation. Tombstone Canyon was a road frequently used by these independent travelers, as it was part of US Highway 80, and businesses opened to meet the needs of people passing through Bisbee. This included providing overnight accommodations for tourists with their own cars, such as the Pueblo Court motel located at the edge of Bisbee on West Boulevard.

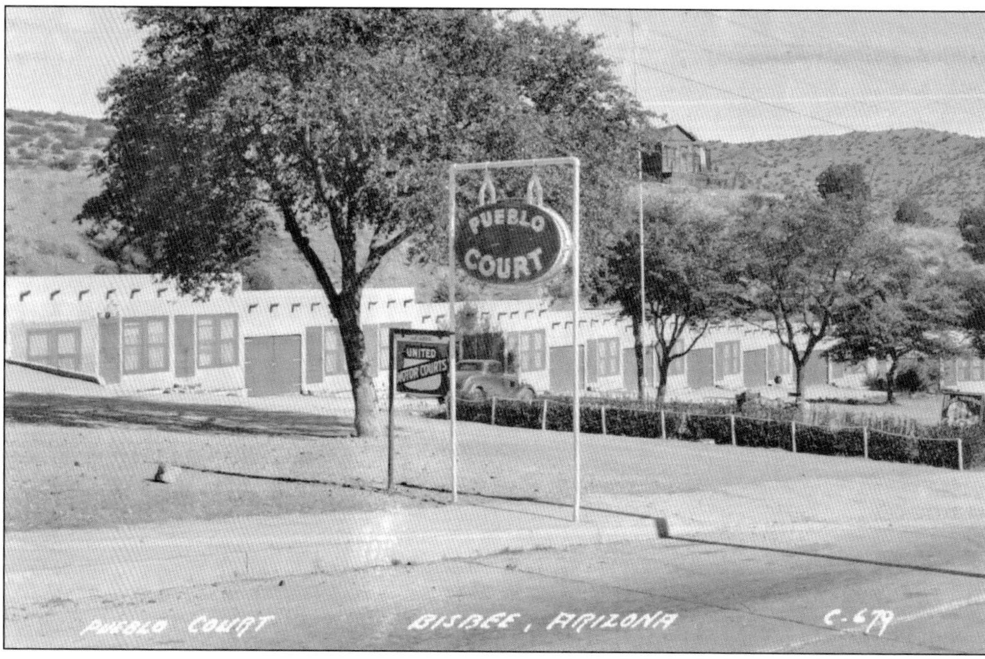

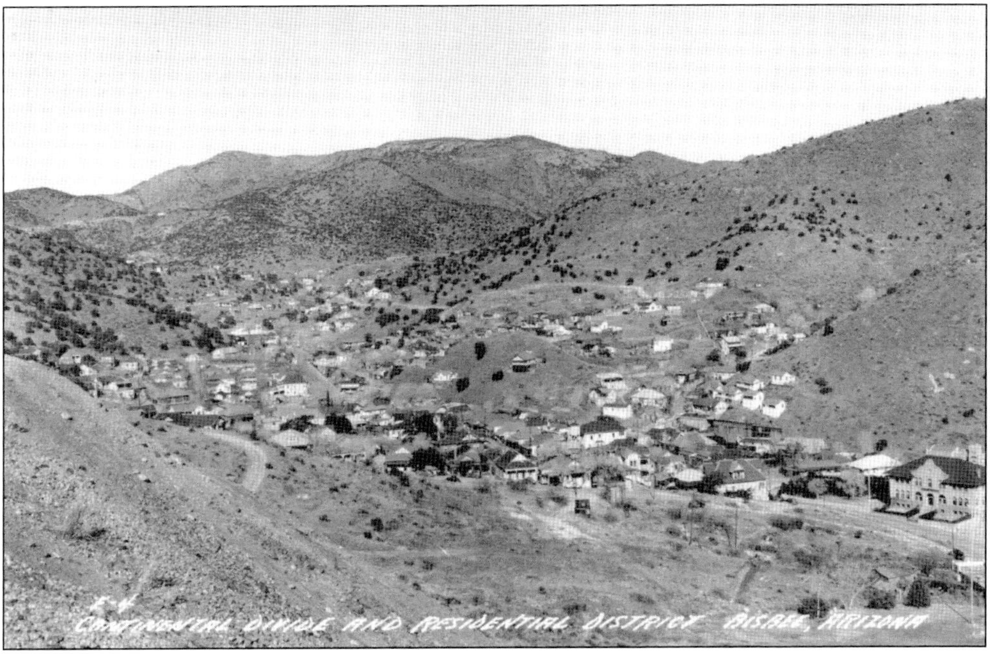

Leaving Bisbee heading west out of Tombstone Canyon, one reaches the mountain pass that separates Bisbee from the outside world. "The Divide," as it is known, is visible in the back left of the 1940s postcard above. Taking the road leading up the Divide offered travelers an amazing view of Tombstone Canyon and the surrounding hills. The 1940s image below reveals the eastern view from the Divide looking over the community. From this vantage point one can not only see Bisbee tucked away in the mountains but also the suburb of Warren in the top left and the mountains of Mexico off in the distance.

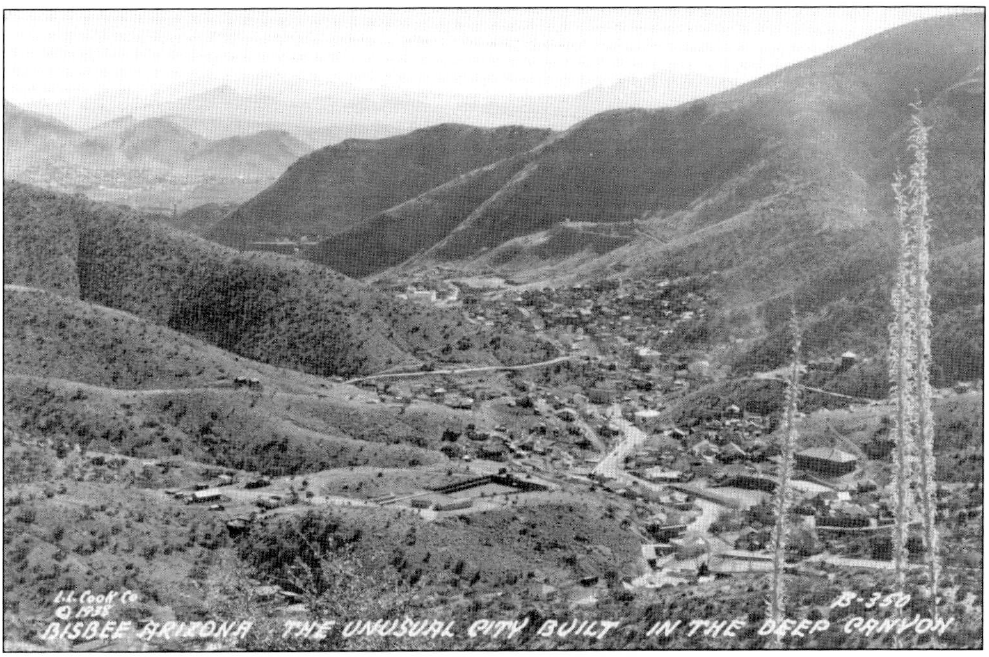

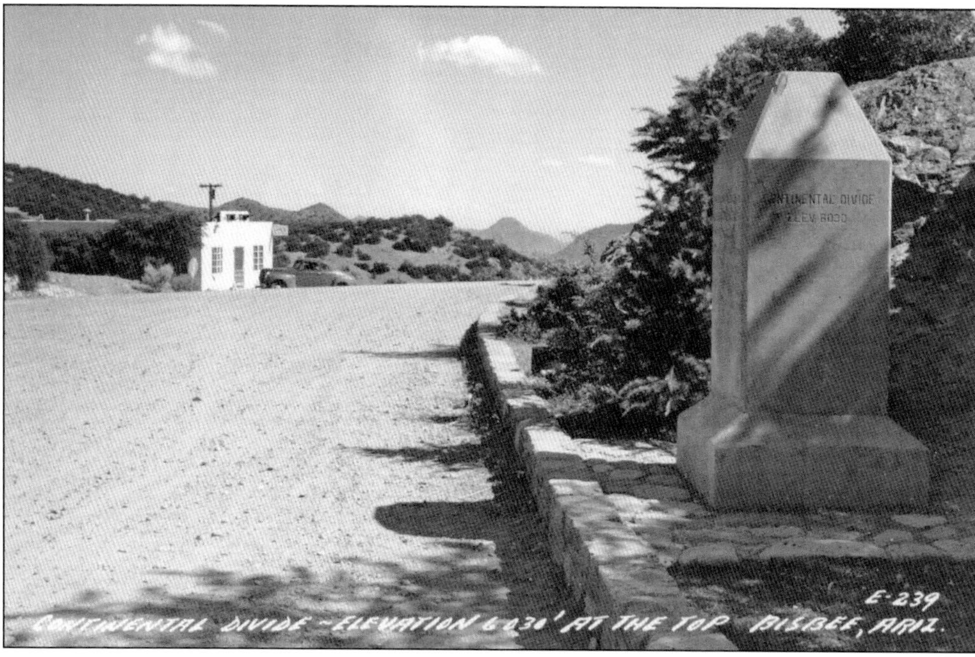

The top of Divide Road, shown above around 1940, reaches an elevation of 6,030 feet. A monument placed at this site incorrectly denotes the area as being part of the Continental Divide, which is located to the east in New Mexico. Divide Road was used as the means of exiting the city to the west, until the Mule Pass Tunnel was created in 1958. After one crested the Divide, the road descended out of the Mule Mountains, enabling travelers to explore the rugged Arizona desert and visit nearby communities.

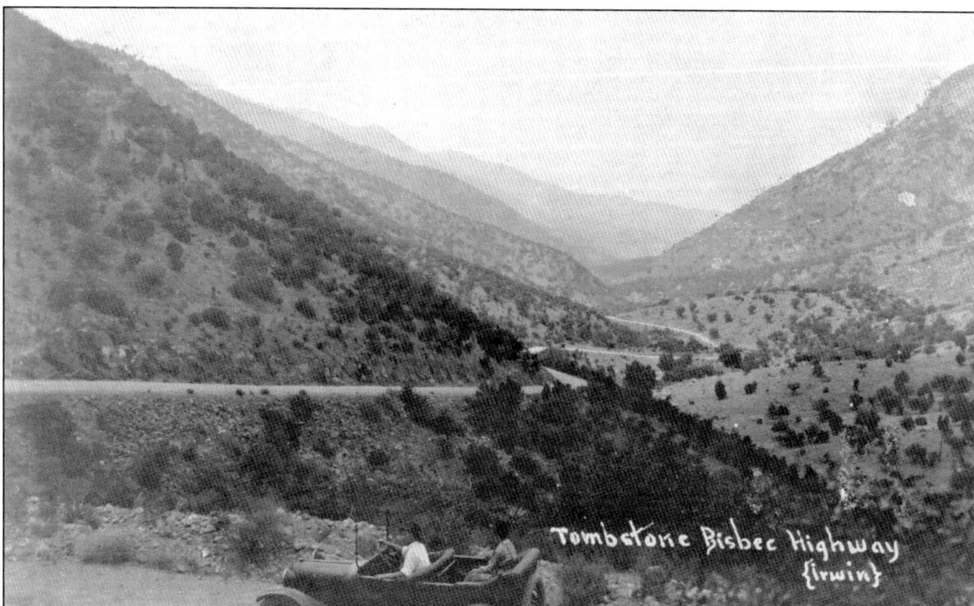

Two

The Queen's Empire

Bisbee is often referred to as the "queen of the copper camps," and with over eight billion pounds of copper extracted from the area in less than 100 years, it is a title rightly earned. Bisbee sits within the Warren Mining District, an area that contained a series of rich copper deposits that spread well beyond the community to the south and east. Evidence of the region's mineral wealth was easily seen on the postcards of Bisbee, as mine headframes clung to the mountainsides, open-pit mines were developed, and nearby piles of extracted rock dotted the desert landscape. To the citizens of Bisbee these were symbols of progress and opportunity. As long as the mines produced ore, they were able to obtain well-paid employment and provide for their families. Such sights were also signs of financial opportunities for investors and a means of marketing projects for mining companies. Early Bisbee photographer W.R. Humphries stated that most mining companies worthy of note used images to promote their properties to investors. "The camera will tell the truth," Humphries stated, and postcards were transportable, visual representations of a mining property hundreds, if not thousands, of miles away. Securing investors was imperative for companies, and though numerous mining ventures sprouted in Bisbee, those properties with the strongest financial prowess ultimately absorbed the smaller businesses.

The rich ore deposits not only offered opportunities to miners and investors, but to a growing country, too. During the early 20th century, the United States was still expanding its borders and adding states to the Union. The mining operations in Bisbee were evidence of the wealth of the Arizona Territory and the financial benefits the United States would receive by making the area a state. Therefore, when postcards depicting Bisbee mining properties were sent to friends and relatives in the East, a clear message was being sent on the front of the card. There are economic opportunities for individuals to be found in Bisbee and mineral wealth in Arizona for a stronger America.

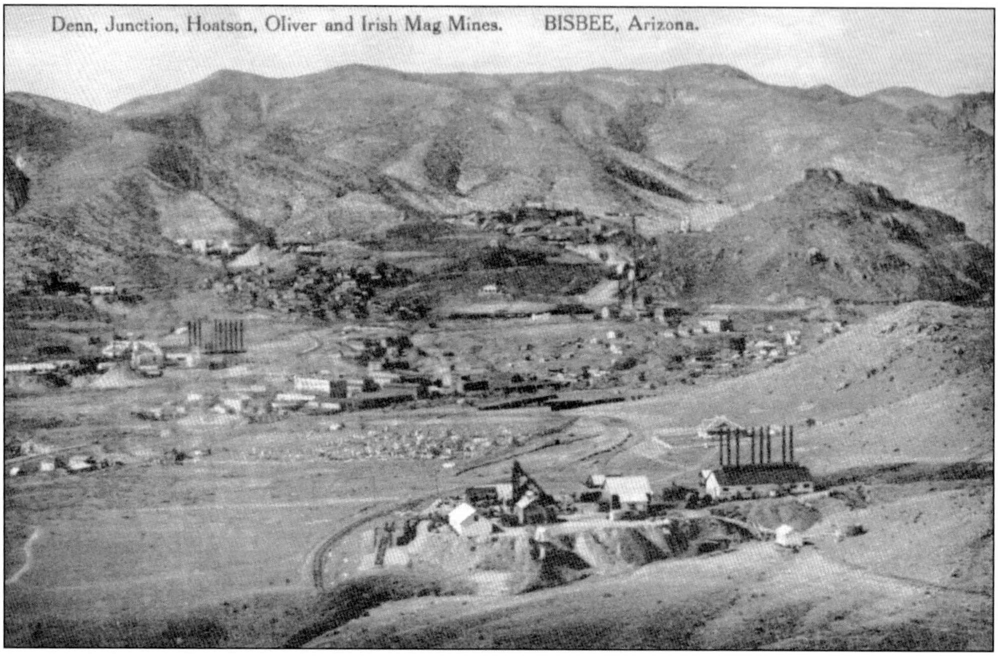

The Warren Mining District boasted over 30 mines. This 1914 postcard shows no less than five operations all within a short distance of each other. This aided both mine exploration and transportation of material from the mines.

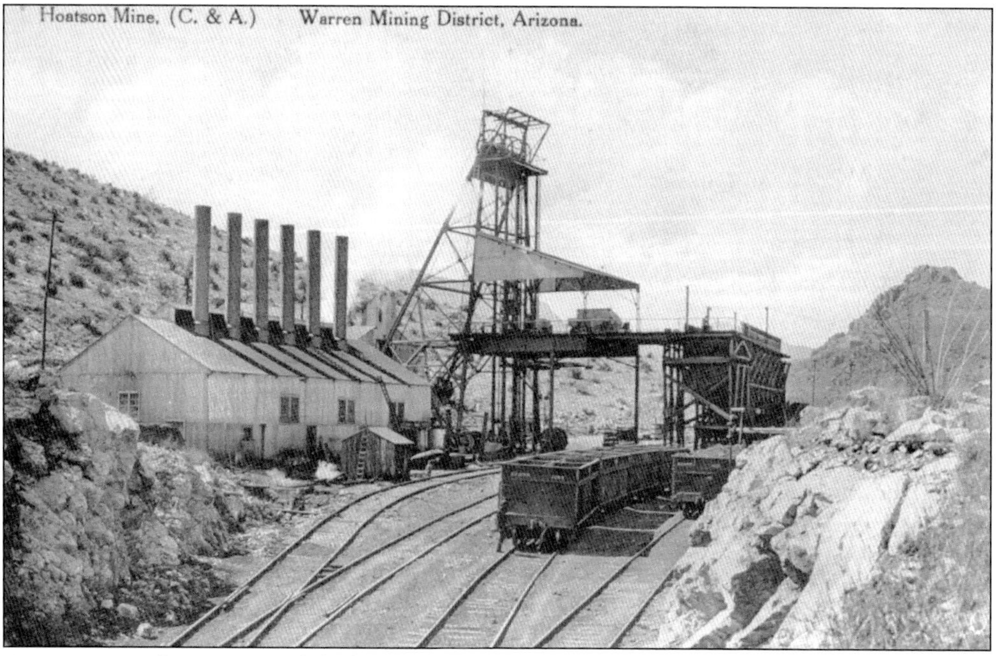

To extract ore from a vertical shaft mine, large structures called headframes were built over the openings, as seen here in 1910. Headframes were made of wood or steel and held sheave wheels that ran cable from a hoist to an elevator that raised ore to the surface. Often, railroad tracks ran next to the structures, facilitating loading of the ore brought to the surface.

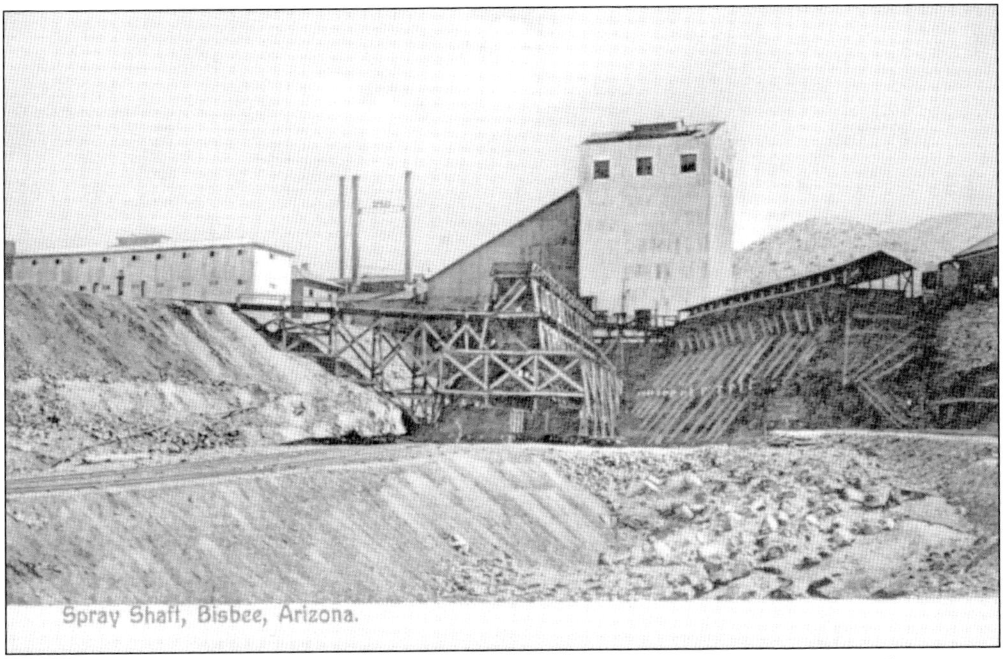

Some mine headframes were enclosed within buildings. This was often done by the Copper Queen Mining Company, as illustrated in this 1910 postcard of the Spray Mine. This was a practice brought from the East where adverse weather posed the risk of slowing production. However, the hospitable climate in Bisbee did not necessitate this undertaking, and the Calumet & Arizona left its headframes unenclosed.

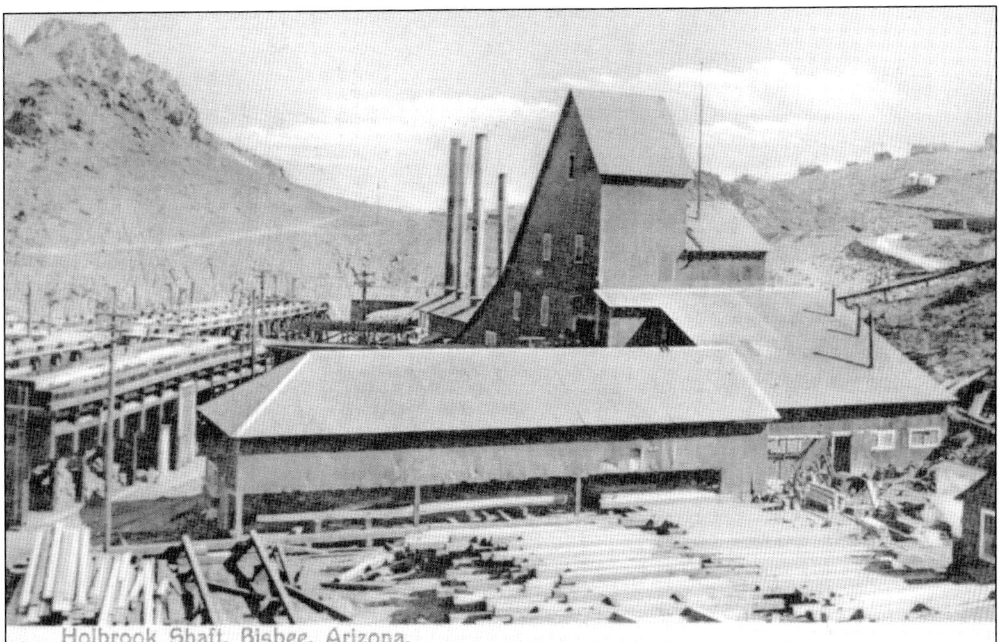

In order to develop the underground mines, timber was needed to support the mine workings. With an astonishing over 2,000 miles of mine workings in Bisbee, millions of trees were needed. This view of the Holbrook Mine about 1910 shows the mine's large timber yard.

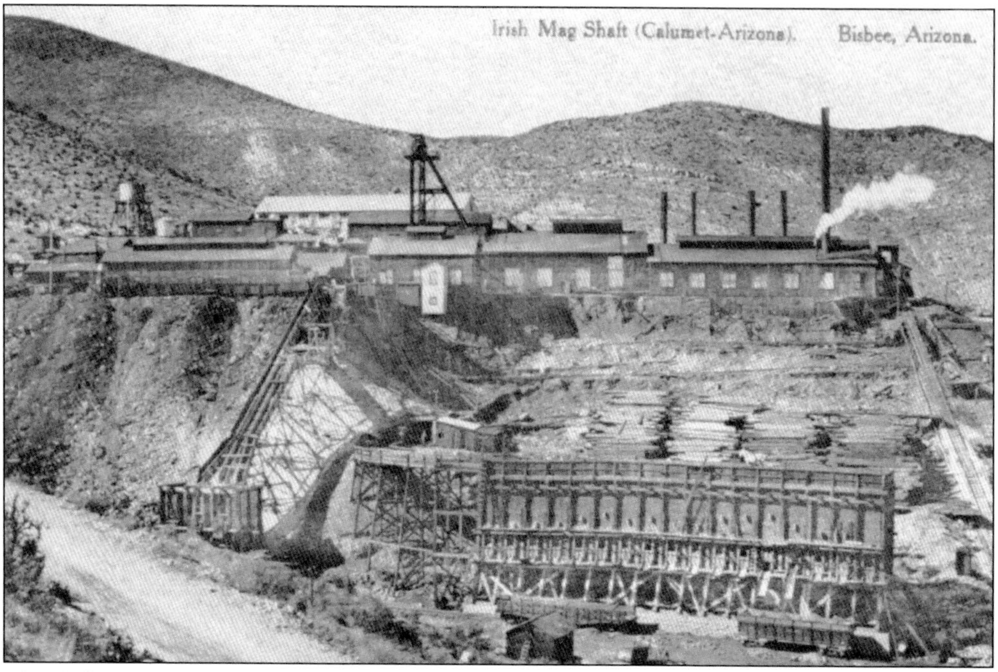

The Calumet & Arizona began development on the Irish Mag Mine in 1900; two years later, the mine yielded an astonishing nine-percent copper ore. Compared to the typical modern average of less than one percent retrieved, the richness of the Warren Mining District's ore deposit is clear. During 11 years of operation, the Irish Mag Mine produced a profit of nearly $10 million.

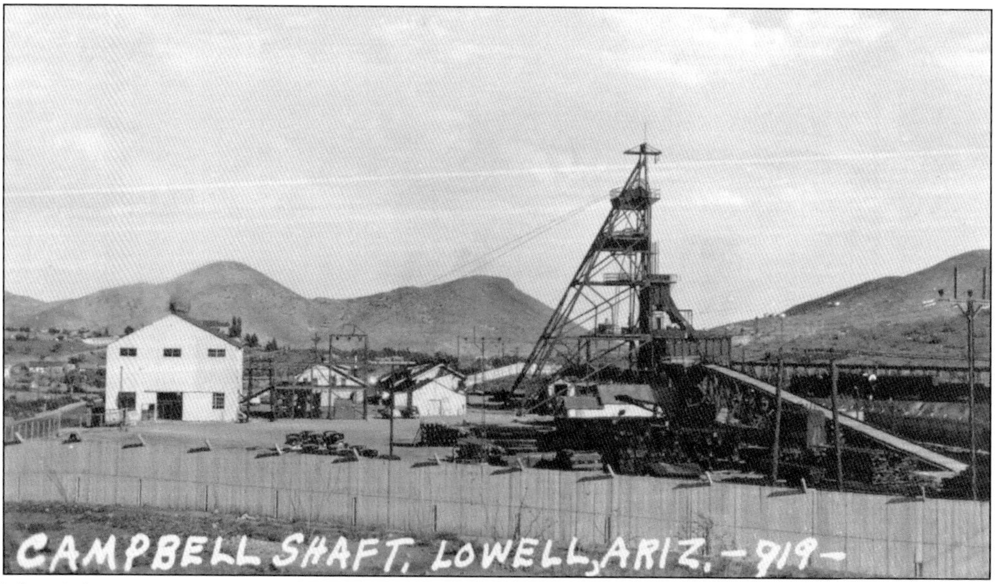

The Calumet & Arizona's Campbell Mine became the last successful vertical shaft mine to be sunk in the mining district, reaching a depth of 3,380 feet. Bisbee's largest ore body located in the district was discovered on the 1,600-foot level of the Campbell Mine. Shown here in 1948, the mine operated until 1975 and produced several million tons of ore and was the most productive in the area's history.

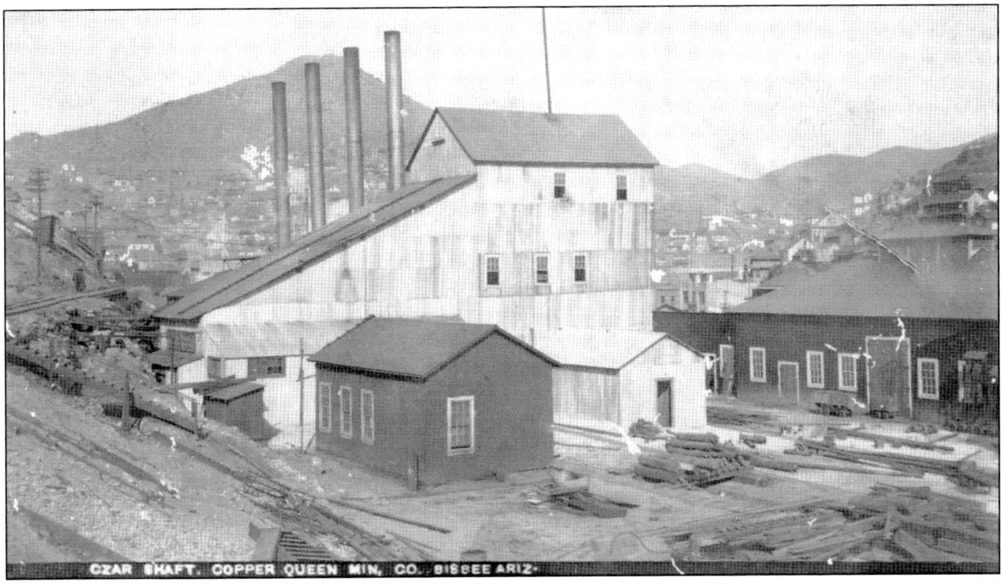

Development of the Czar Mine began in 1884. It went on to be one of the most successful mines for the Copper Queen Mining Company. It was the longest operating mine, closing in 1944. The parking lot for Bisbee's tourist attraction, the Queen Mine Tour, sits where the Czar Mine's headframe once stood. A silver cap in the parking lot denotes the location of the shaft.

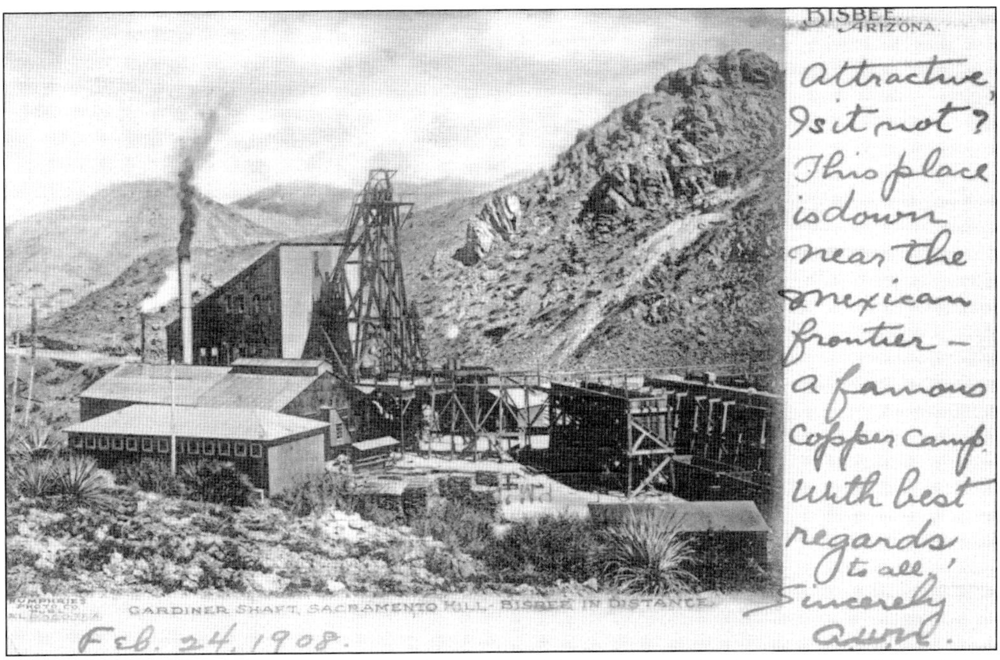

The mines in Bisbee were the richest in the Arizona Territory, and the importance of the mining camp was conveyed throughout the world in both words and images thanks to postcards. This 1908 postcard of the Copper Queen Mining Company's Gardner Mine reads, "Attractive, is it not? This place is down near the Mexican frontier—a famous copper camp."

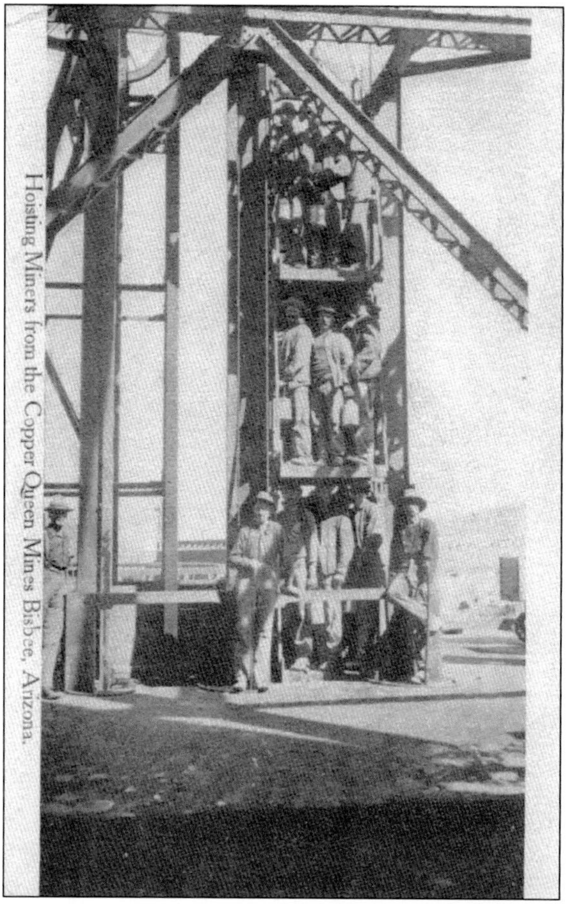

In addition to moving ore and supplies, headframes had elevators that transported miners underground. This 1910 picture of the Gardner Mine illustrates how the men used the shaft elevators, with three levels.

As miners prepared to make their way underground, they gathered the supplies needed for their shift. Most importantly, this included light to work with, as seen by candles held in the hands of these miners during 1906. This was their primary light source in the early years of mining. Note the man standing on the left, whose pant leg is covered in white candle wax.

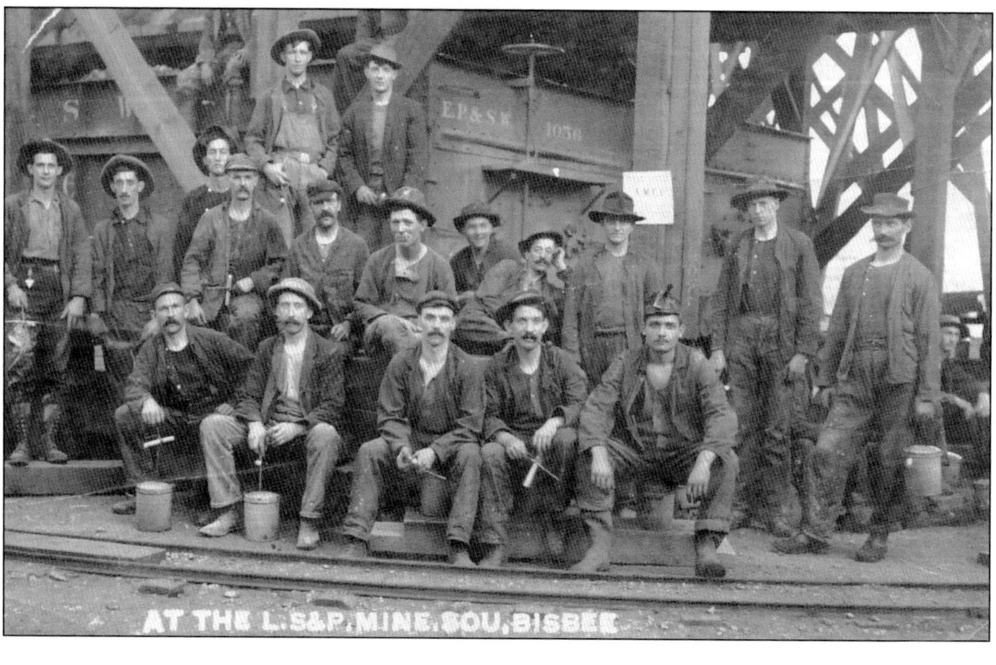

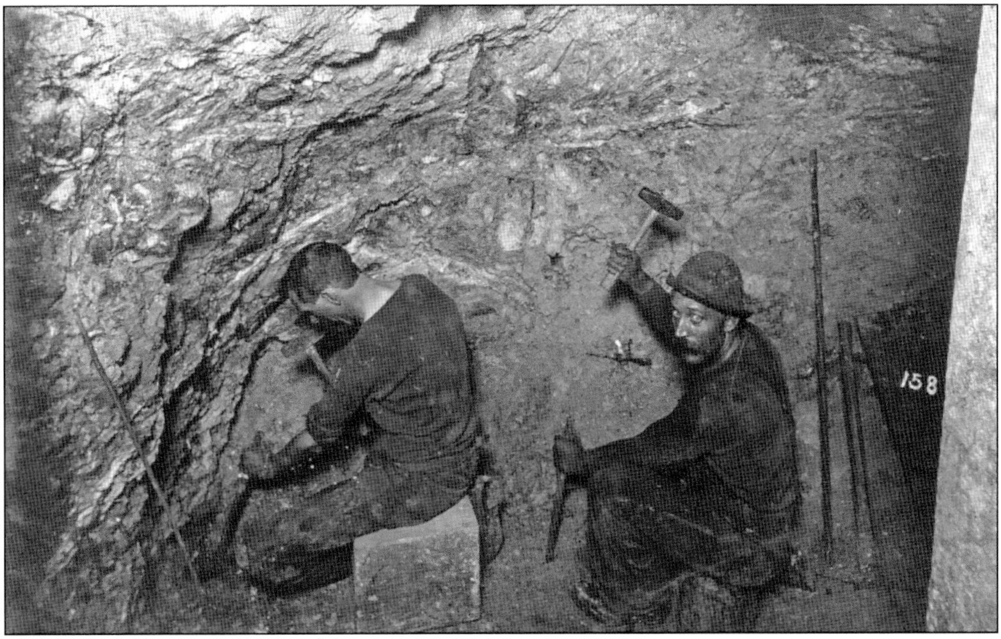

To remove the copper ore from the underground mine, early miners used chisels and hammers to make blast holes, creating a number of small round openings up to five feet deep. This process is demonstrated here in 1904. The miners then inserted sticks of dynamite in the openings. The explosion loosened rock, which would be shoveled into ore cars for its journey to the surface.

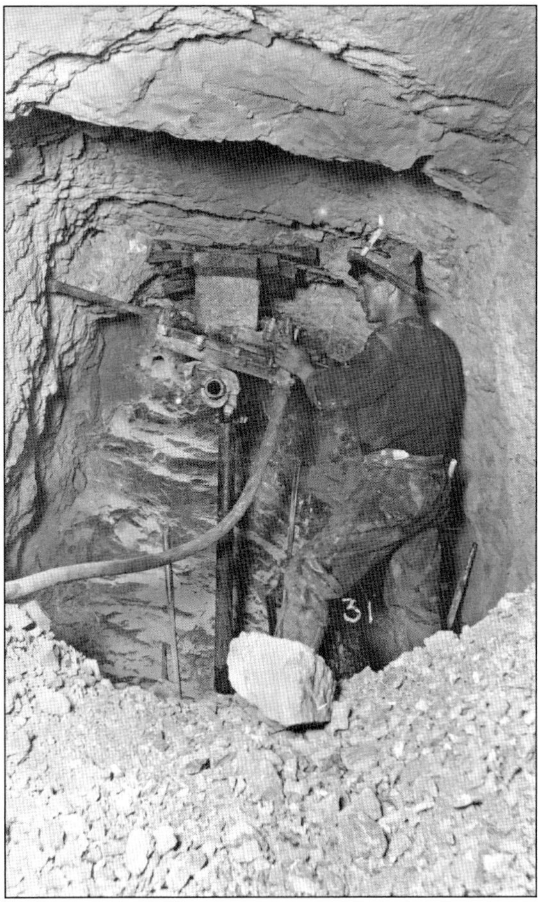

By 1908, advances in mining technology brought the pneumatic drill into mines across the country. This enabled a miner to do the same amount of work in hours that would take days using a traditional hand drill. An air hose leading to the machine helped clear the drill hole. However, the dust created was inhaled by miners, causing health problems for the men.

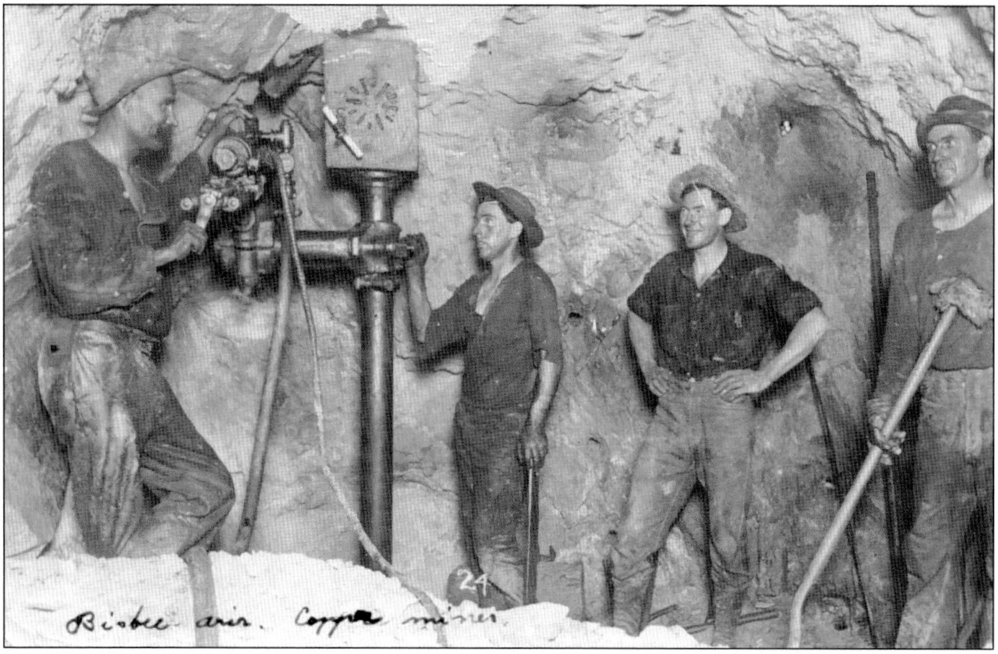

The introduction of water into the drilling process reduced the problem of rock dust that plagued earlier drills. This was supplied by a second hose that sent water down a drill steel and sprayed out of the bit, eliminating the dust released into the air. An example of this style of modified drill is seen in this 1911 image.

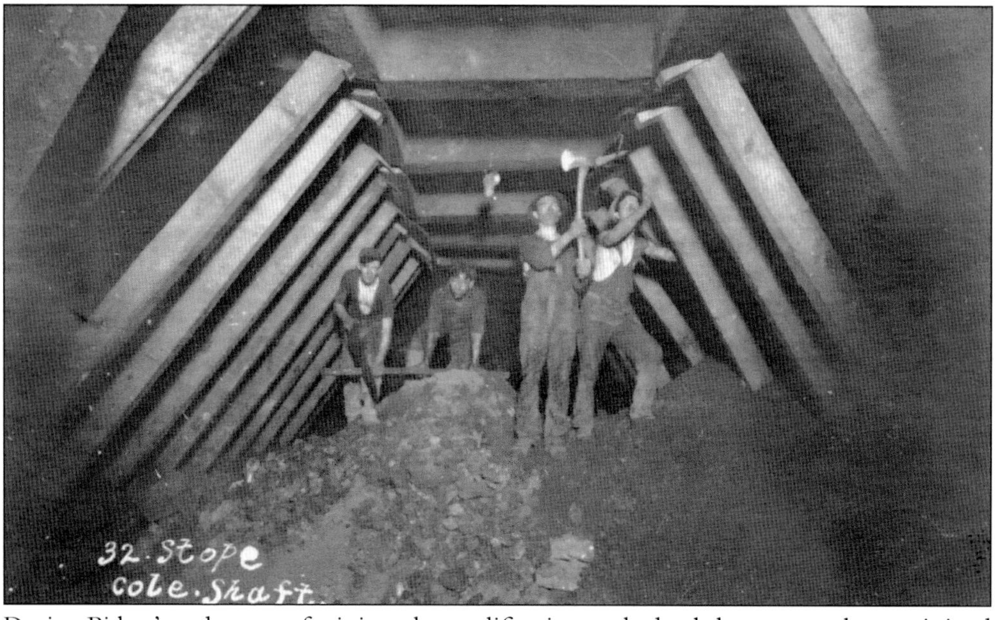

During Bisbee's early years of mining, the modification to the land above ground was minimal compared to what was occurring beneath the surface. The talented miners of Bisbee developed a new world hundreds of feet below the ground, one that grew larger as they followed the rich copper ore deposited in the rock. This 1916 postcard shows miners placing supporting timbers where they have removed ore.

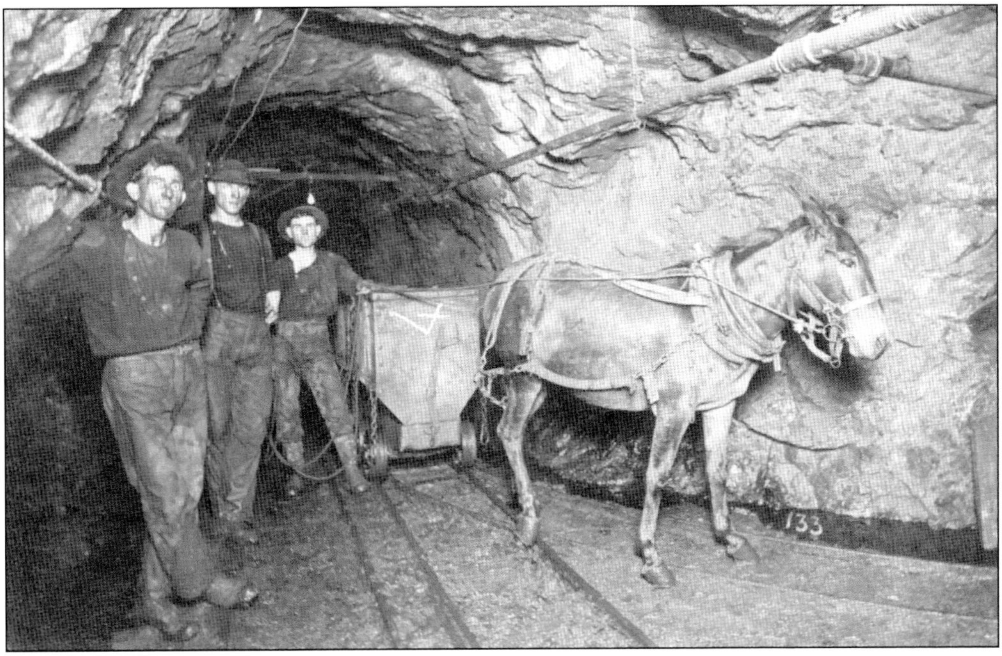

Beginning in 1907, mules were brought underground to pull heavy loads of ore. The introduction of mules enabled three to five ore cars full of material to be hauled, whereas the miners were only capable of pushing a single car.

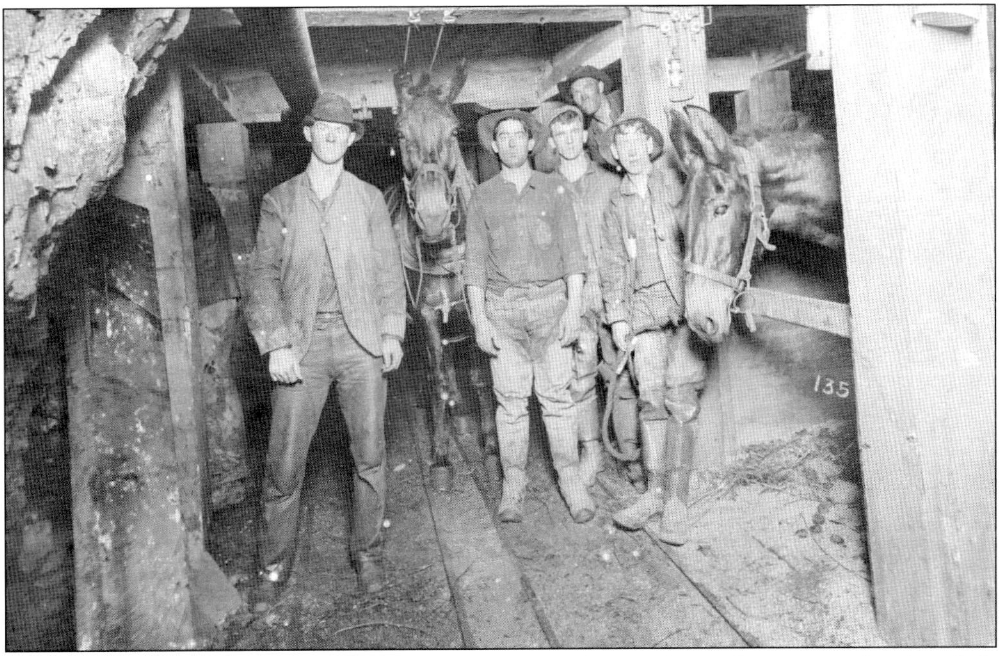

This rare image of a mule barn shows where the animals were kept when not performing their invaluable labor. The animals spent years underground, but were well taken care of, stabled in dry, well lit, and warm areas.

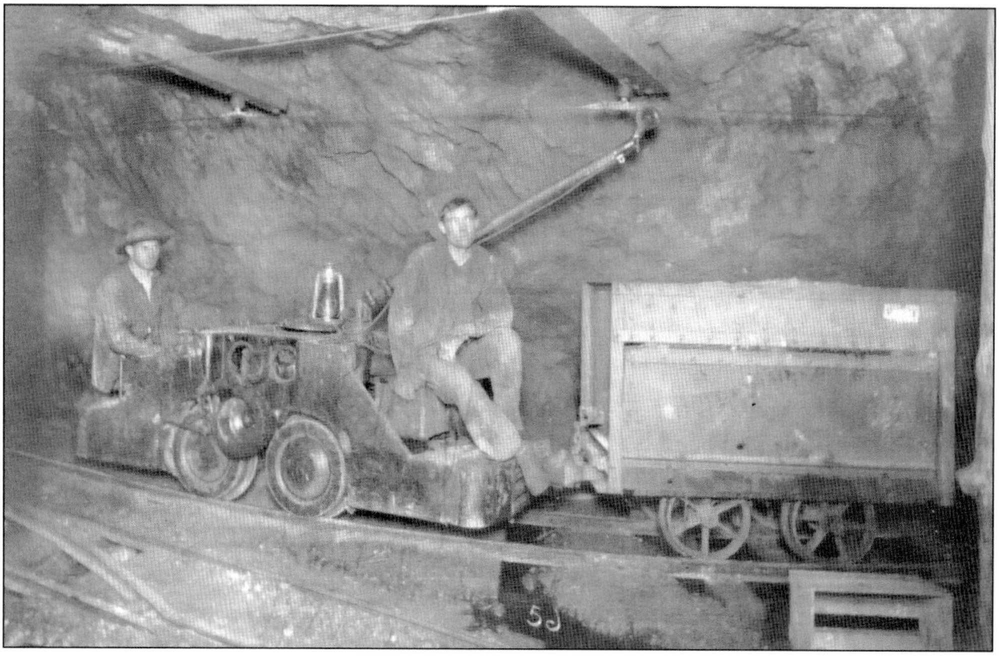

By 1909, advances in the mining industry began replacing mules with trolley motors to transport the copper ore. The motors operated on 240 volts of direct current electricity that ran through the bare copper wire visible overhead in both these images. This new method of transportation significantly increased volume, as it hauled eight to ten ore cars, doubling the load pulled by mules.

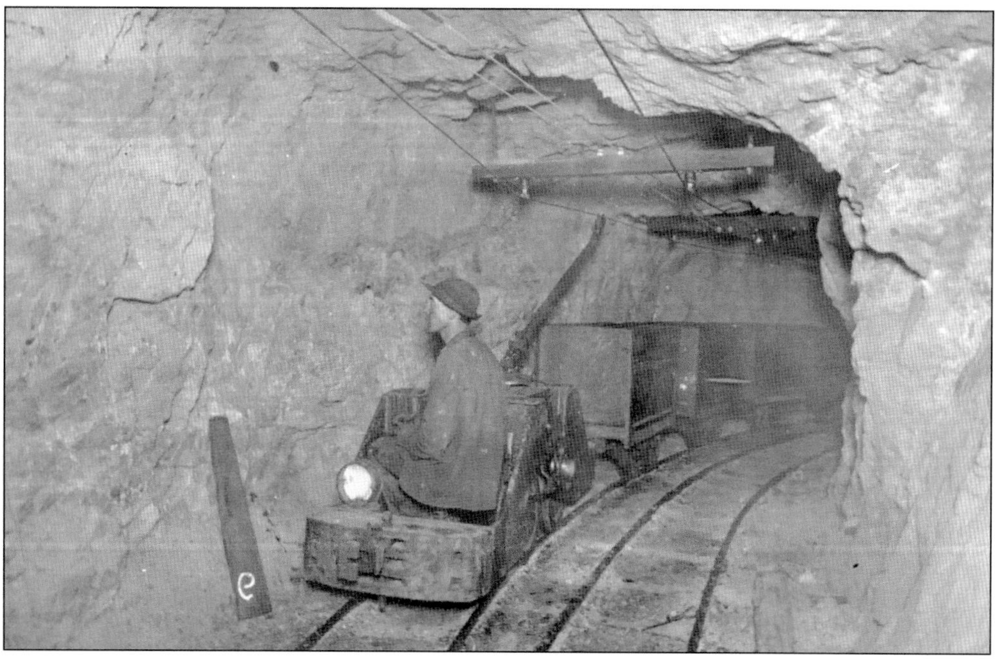

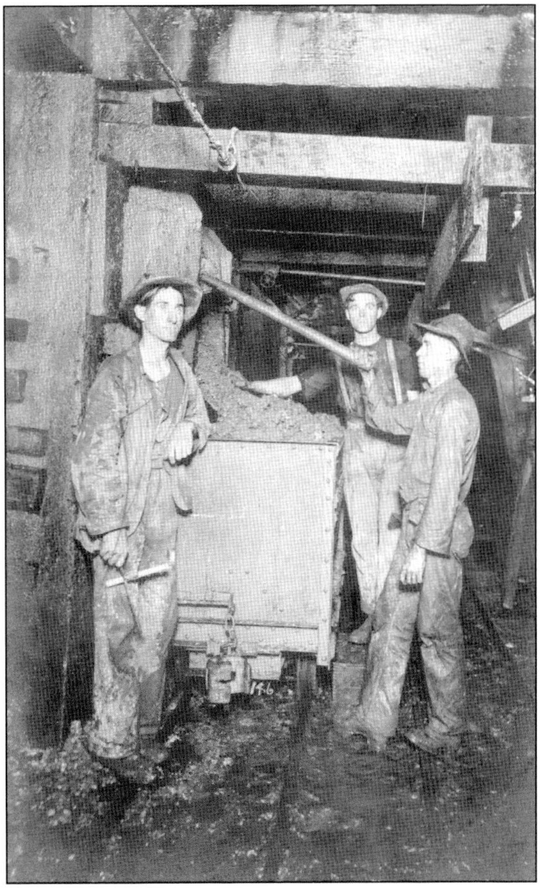

Chutes were used to load material into mine cars for transporting the ore to the surface. At right, a man is pushing down on a bar that opened the chute door and released the ore into the cars. Once they finished traversing the tunnels within the mines, the ore-filled cars arrived at the shaft and were taken to the surface using shaft elevators. This is illustrated in the 1910 card below. These elevators carried miners, ore, and timber and other supplies.

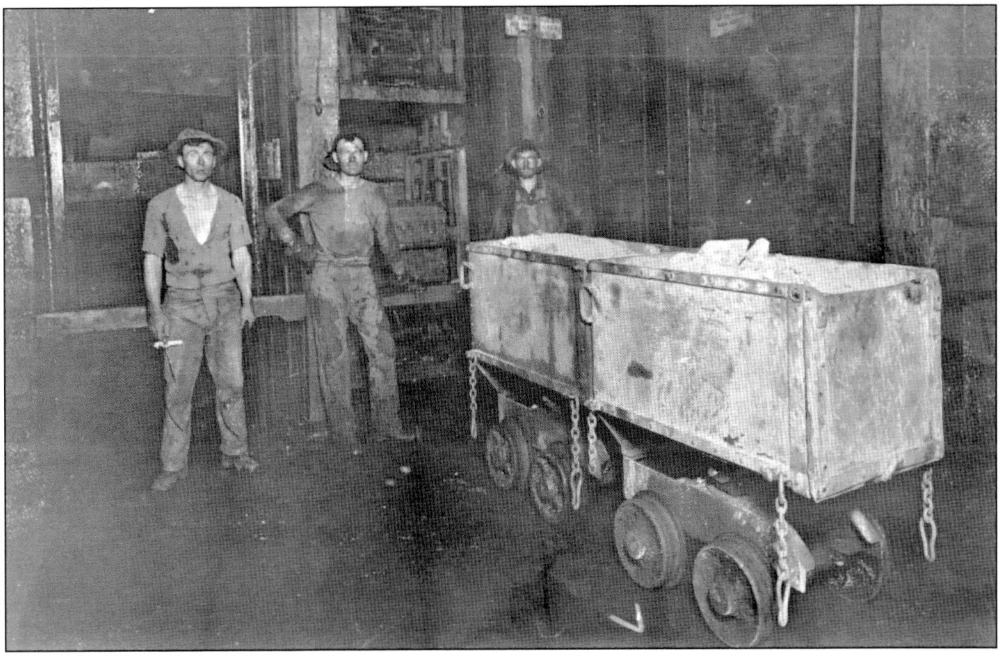

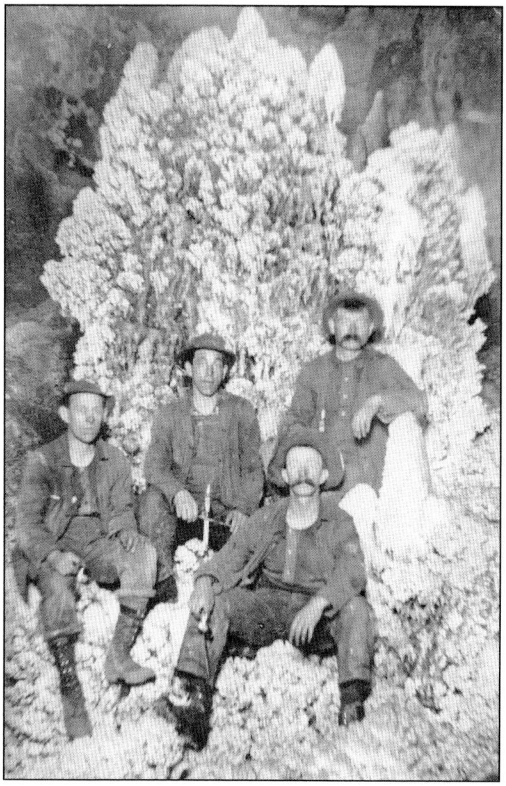

Occasionally during the miners' quest for ore, they would find another world on the other side of the rock—crystal caves. Hundreds of these caves were discovered in Bisbee, many filled with brightly colored ore minerals. The Shattuck Cave was one of Bisbee's best known, pictured here in 1913. The beauty of these spaces was not lost to the miners or Bisbeeites. The caves were used to hold special events such as dances and were meeting sites for fraternal organizations. While appreciated by the miners, their jobs still had to be accomplished and most of these caves were filled in with waste rock or collapsed due to an unstable environment caused by mining in the proximity. (Both, courtesy of the Bisbee Mining & Historical Museum, Metz Collection.)

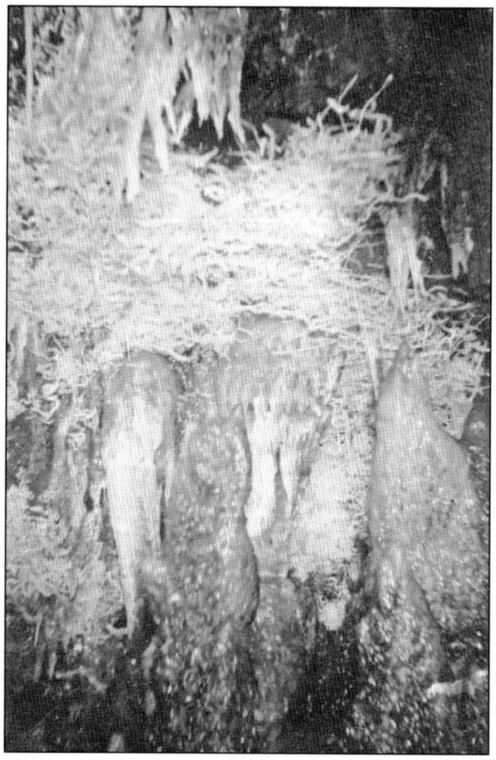

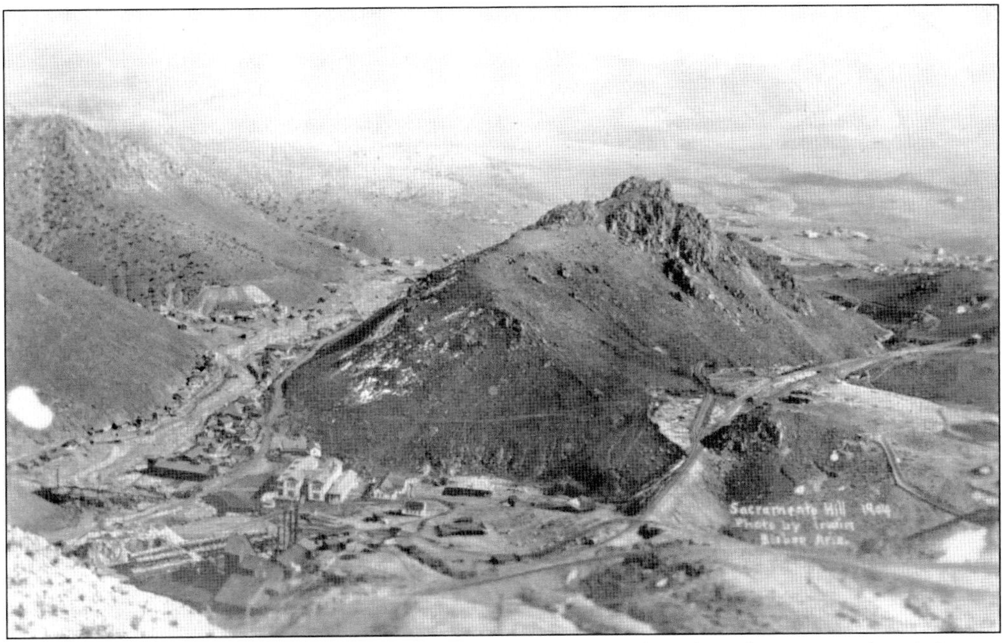

In addition to the underground, companies in Bisbee employed surface open-pit mining to extract ore. This postcard is a 1904 view of Sacramento Hill, located southeast of downtown Bisbee. The area became the location of the community's first surface open-pit mine in 1917 when the development of the Sacramento Pit began.

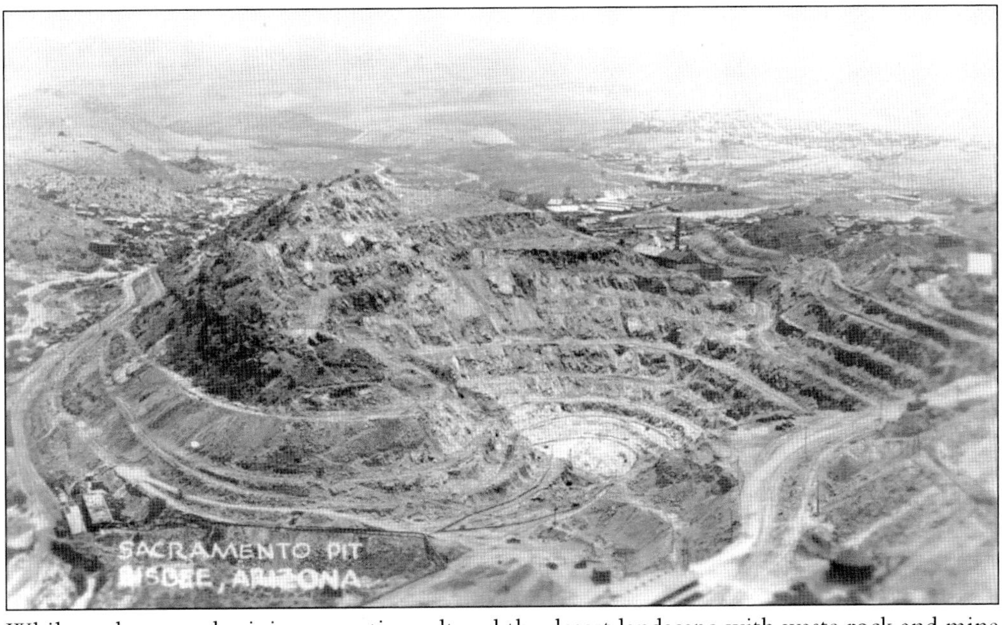

While underground mining operations altered the desert landscape with waste rock and mine workings, the creation of an open pit dramatically changed the look of Bisbee. Pictured here in about 1930, a significant portion of Sacramento Hill has been excavated with less than half of the mound remaining. The Sacramento Pit was one of the earliest open-pit mining operations in the United States.

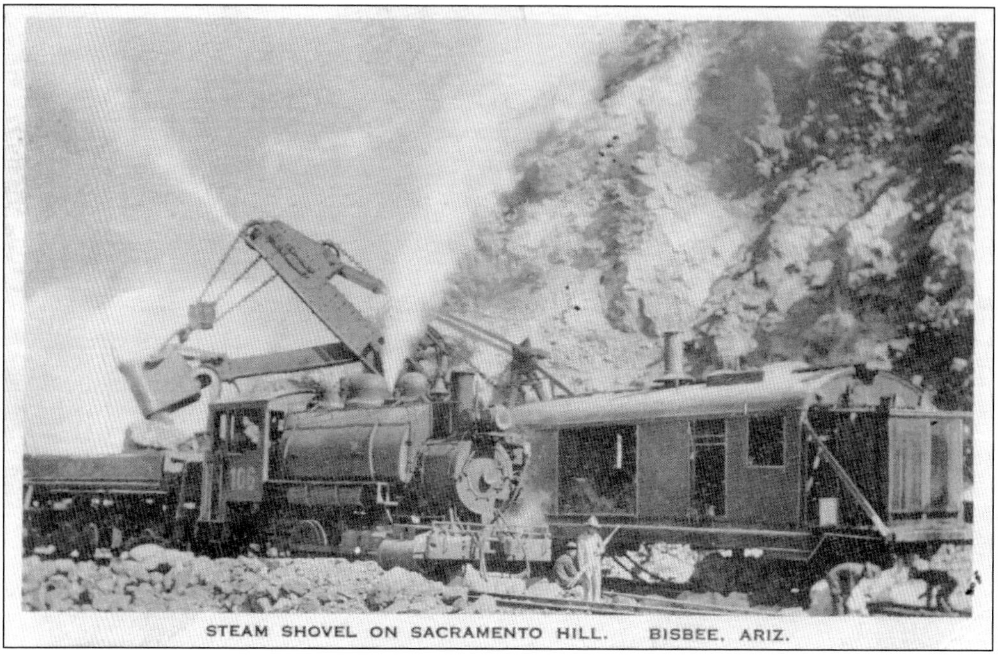

STEAM SHOVEL ON SACRAMENTO HILL. BISBEE, ARIZ.

Open-pit operations took the basic techniques of mining and used them on a large scale, illustrated in these 1919 (above) and 1922 (below) postcards. The removal and transportation of ore was now carried out by massive machines, which performed the same amount of work in mere moments that it took men days to perform. An explosion loosened material, enabling the steam shovels to scoop up rock and dirt and load it into large train cars. The train would then transport the material to begin processing at the concentrator.

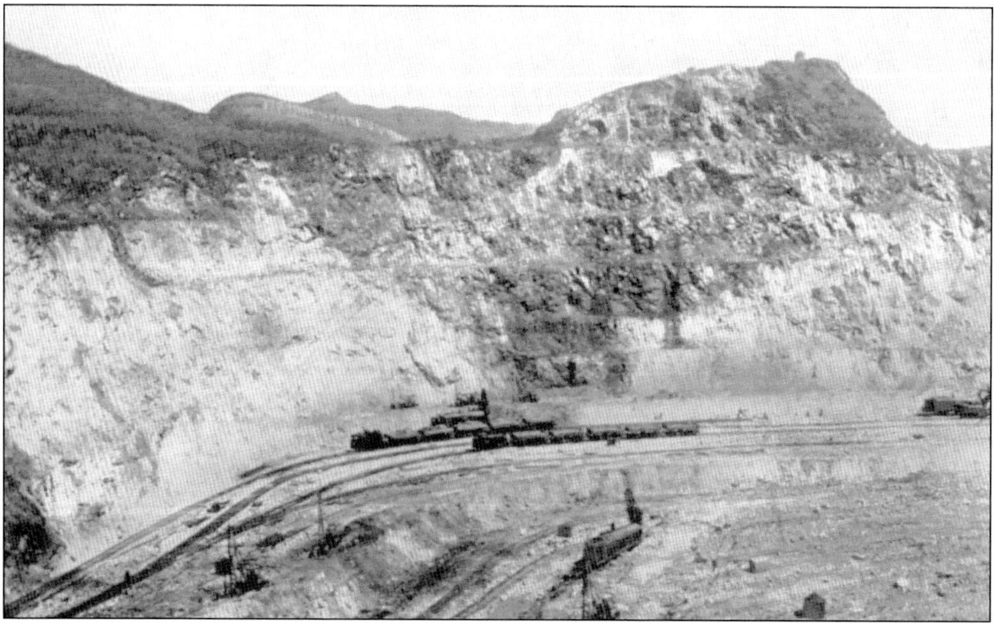

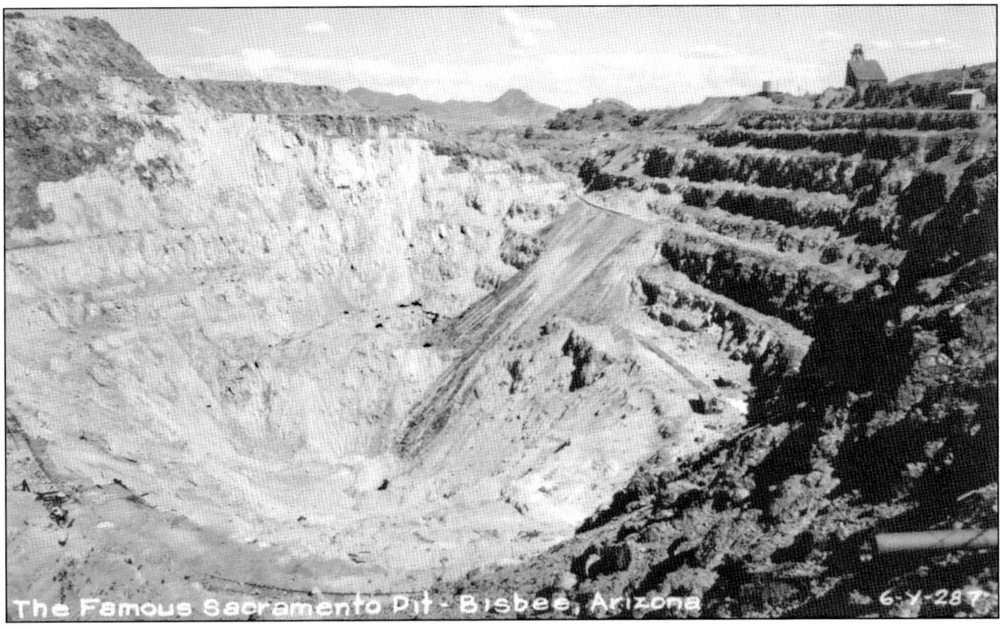

The Sacramento Pit reached a depth of 700 feet, with 32,800,000 tons of material being removed from the ground. Of this, nine million was processed with 23 million classified as waste material. By 1938, improvements in mining technology enabled companies to revisit old pits and waste piles to make money from small percentages of copper ore in material that companies formerly could not financially feasibly extract.

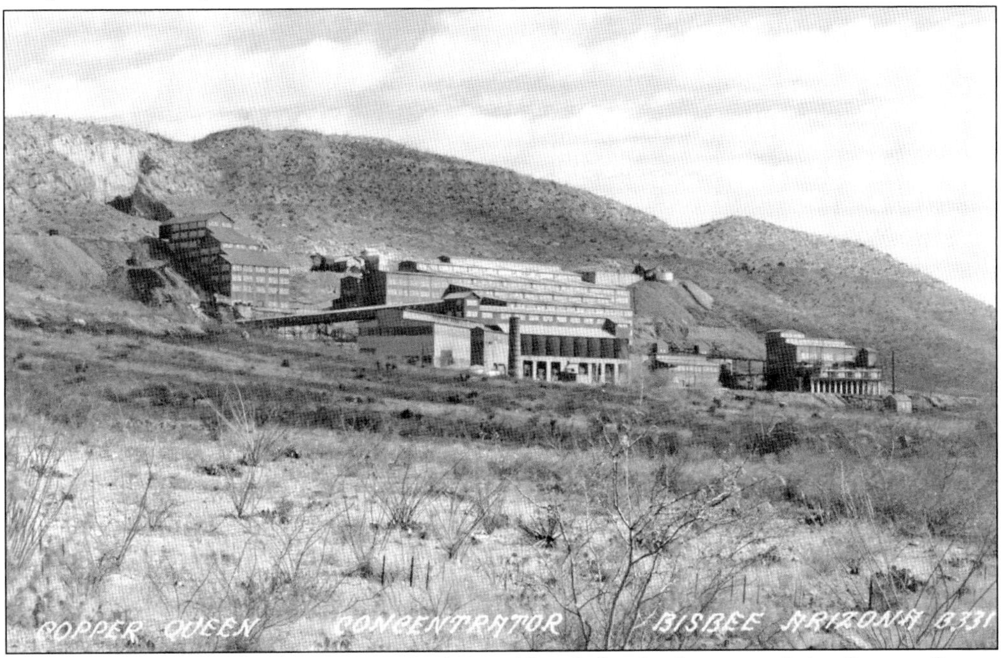

In order to make mining Bisbee's ore bodies viable, Phelps Dodge erected the Copper Queen Concentrator near the settlement of Tovreaville in 1923. The concentrator separated the waste material from the copper ore, reducing the time and transportation costs of having to process the waste elsewhere.

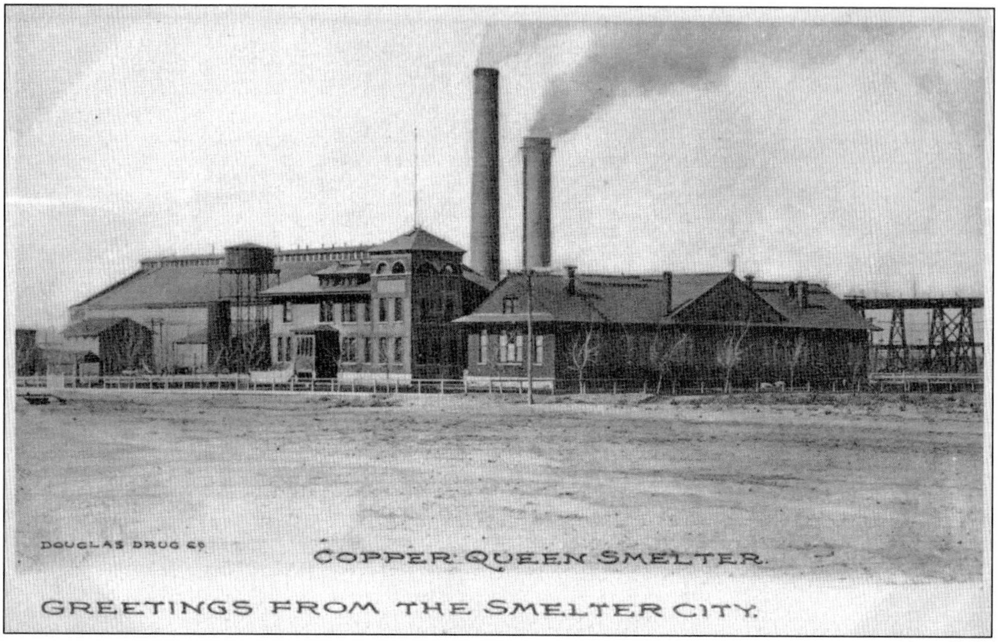

By 1904, both the Copper Queen and Calumet & Arizona mining companies transported ore to a new smelter in nearby Douglas for processing. Here, the facilities are pictured in 1904 (above) and 1930 (below). Before this, a smelter was located in Bisbee, but the smoke it released negatively impacted the air quality in the community, with the surrounding Mule Mountains trapping the pollution inside the canyons. In an effort to improve the air quality in Bisbee, the mining company relocated smelter operations to Douglas. Along the Mexican border to the east of Bisbee, Douglas was an ideal location for the smelter as it was situated in an open valley that enabled the smoke to blow away with the wind.

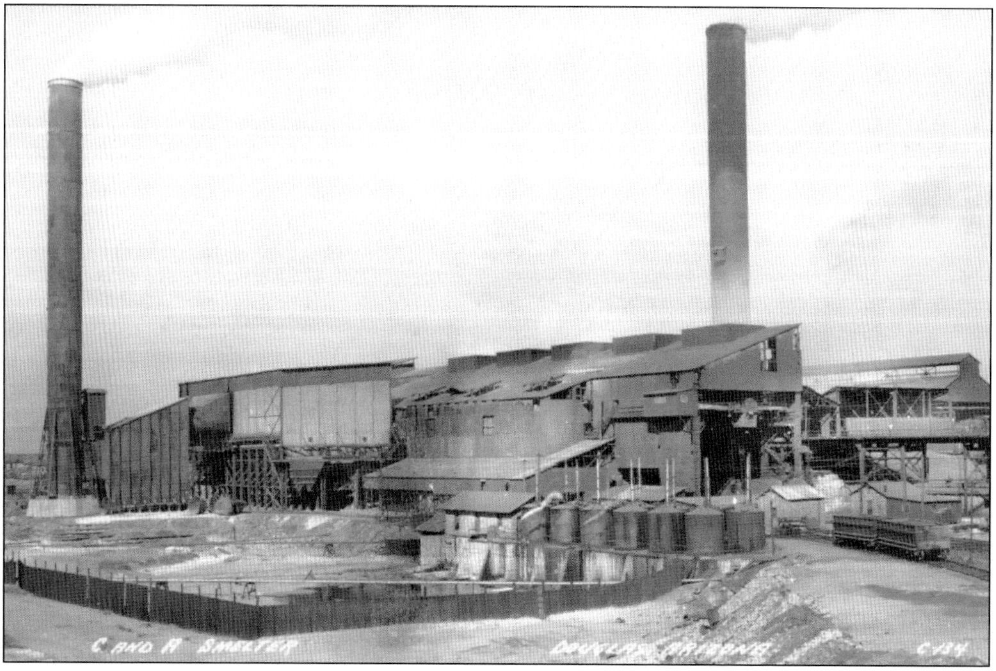

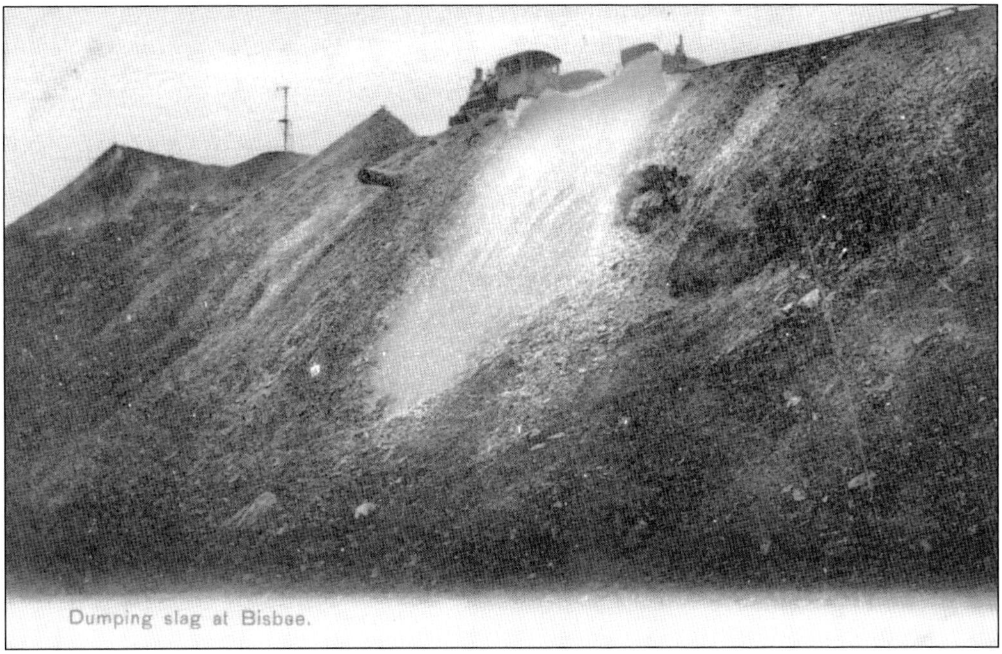

The smelting process involved copper being separated from the impurities in the ore by being exposed to extreme heat. This resulted in the copper, as well as the other material, transforming into a molten state. The copper was poured and cooled in molds while the waste material, known as slag, was poured onto the dump piles. This process was occasionally done in the evening, providing onlookers with a beautiful display. Residents of Bisbee and Douglas fondly recall watching the bright-red molten material being poured upon the dump. The stunning event also caught the eye of photographers, as they captured the scene for the local postcard market.

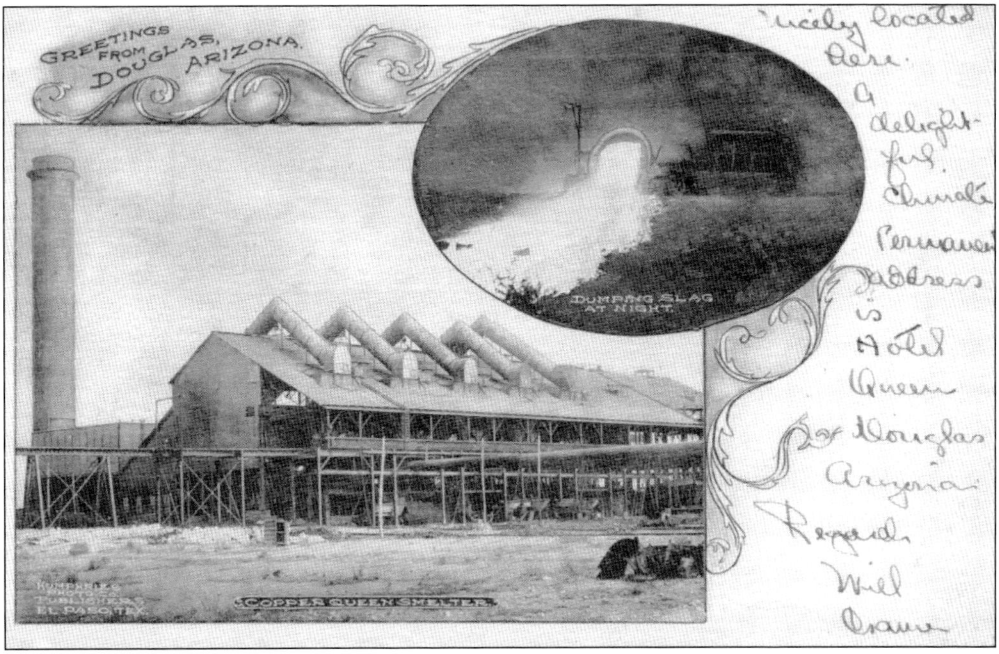

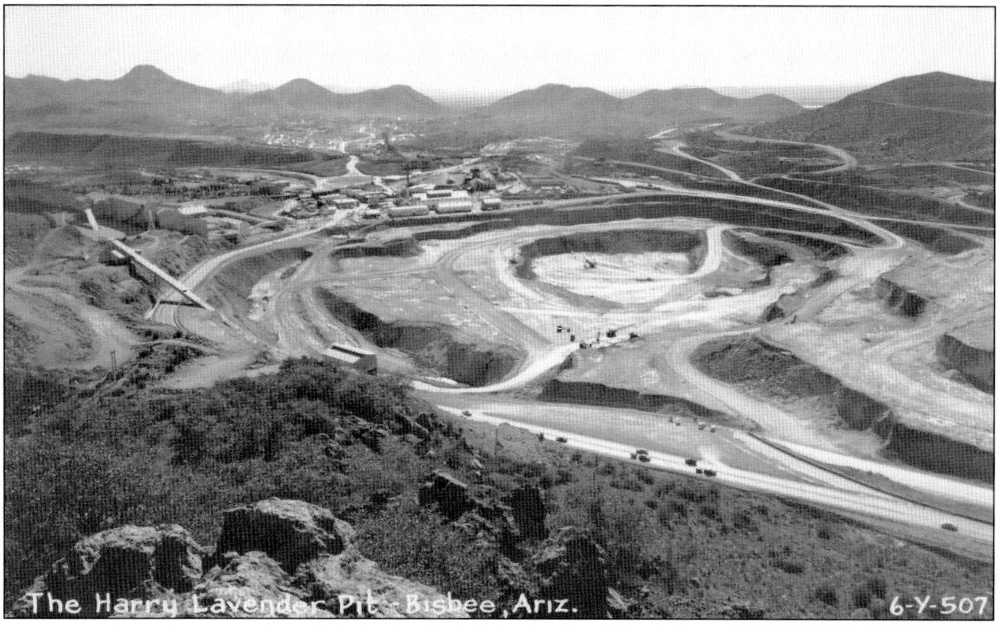

Development of Bisbee's second open-pit mine, the Lavender Pit, began in 1950. This was a time when the US government actively sought strategic metals to make supplies for the Korean War. To encourage the production of copper in Bisbee, the government agreed to buy 112,500 tons of copper produced from the Lavender Pit from the mining company. The Lavender Pit was a successful venture, producing 94,400,000 tons of ore from 1954 until the mine closed in 1974. However, the mine caused the most dramatic change to the landscape the city would ever witness. Many small suburbs were located on top of the pit's ore deposit and had to be moved in order to develop the mine.

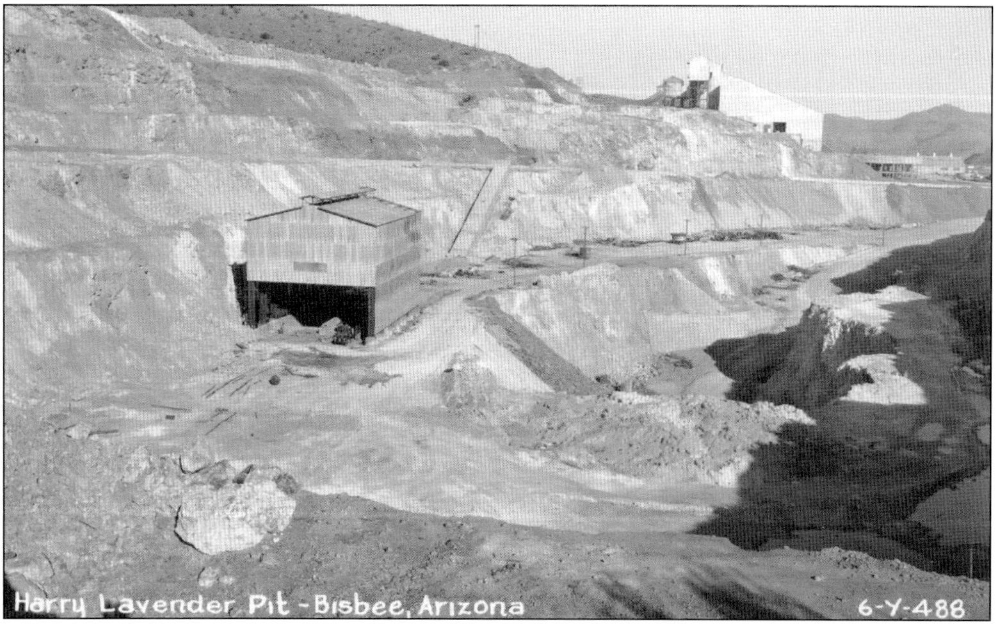

Three

New Horizons

As miners continued to follow the area's scattered ore bodies in the Warren Mining District, new mine sites were established to the south and east of Bisbee during the late 1800s and early 1900s. These new properties provided employment opportunities for thousands of miners in the region. The primary method of transportation to work for these employees was walking, as these mines were developed before the rise of the automobile and public transportation in the Bisbee area. Consequently, communities sprouted up around the mines, including Lowell, Jiggerville, and Johnson Addition, with businesses soon following that provided goods and services to the area's residents. However, Bisbee's rich ore bodies proved to be both a blessing and a curse for residents of these communities. During the 1950s, Phelps Dodge wanted to expand its operations to include a new open-pit mine, the Lavender Pit. Johnson Addition, Lowell, and Jiggerville sat on the areas the mining company wanted to develop, and the vibrant communities that were once home to families and businesses disappeared. In 1973, the *Brewery Gulch Gazette* reminded readers that "some old-timers may have regrets about [the] expansion, but in a mining camp you can't spend much time looking back—you spend it looking forward and in finding and developing additional ore reserves. For it is the precious copper and other metals coming out of the earth that gives life to the Bisbee District, the smelter town of Douglas, all of Cochise County, and to the state and the nation."

While these sections of the mining district vanished, other permanent communities were formed, such as Warren. Warren was developed by the Calumet & Arizona Mining Company in 1906 in an effort to provide an environment that would attract and retain the finest, most dependable employees available. These were typically men who had responsibilities, such as families to provide for, and the Calumet & Arizona designed a community with these individuals in mind. Warren offered everything a family could desire, including educational and recreational facilities and homes to purchase with modern amenities.

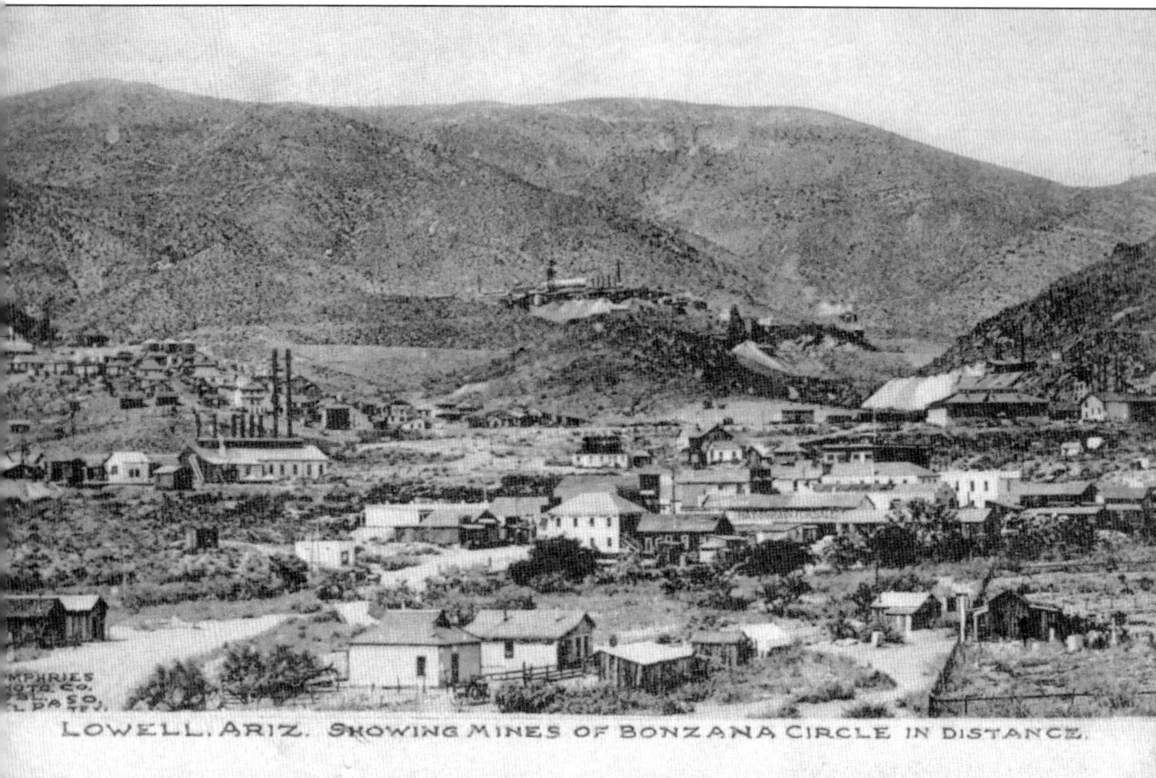

As the potential of the Warren Mining District's ore body continued to be realized, more mines began to be developed around the periphery and to the east of Sacramento Hill. The mines in this region primarily belonged to the Calumet & Arizona and the area became known as the Bonanza Circle. This included properties such as the Lowell, Sacramento, Junction, Hoatson, Oliver, Irish Mag, and Gardner Mines, many of which are shown in this 1902 postcard. The mining company leased land around these sites to miners so they could build homes and have a short distance to walk to work. Soon, the settlement of Lowell was born, taking its name from the nearby shaft. Businesses were also established in Lowell, providing individuals access to goods and services without having to travel to Bisbee. Most of these stores were located on Erie Street, shown in the center of the image, which was the area's main commercial thoroughfare.

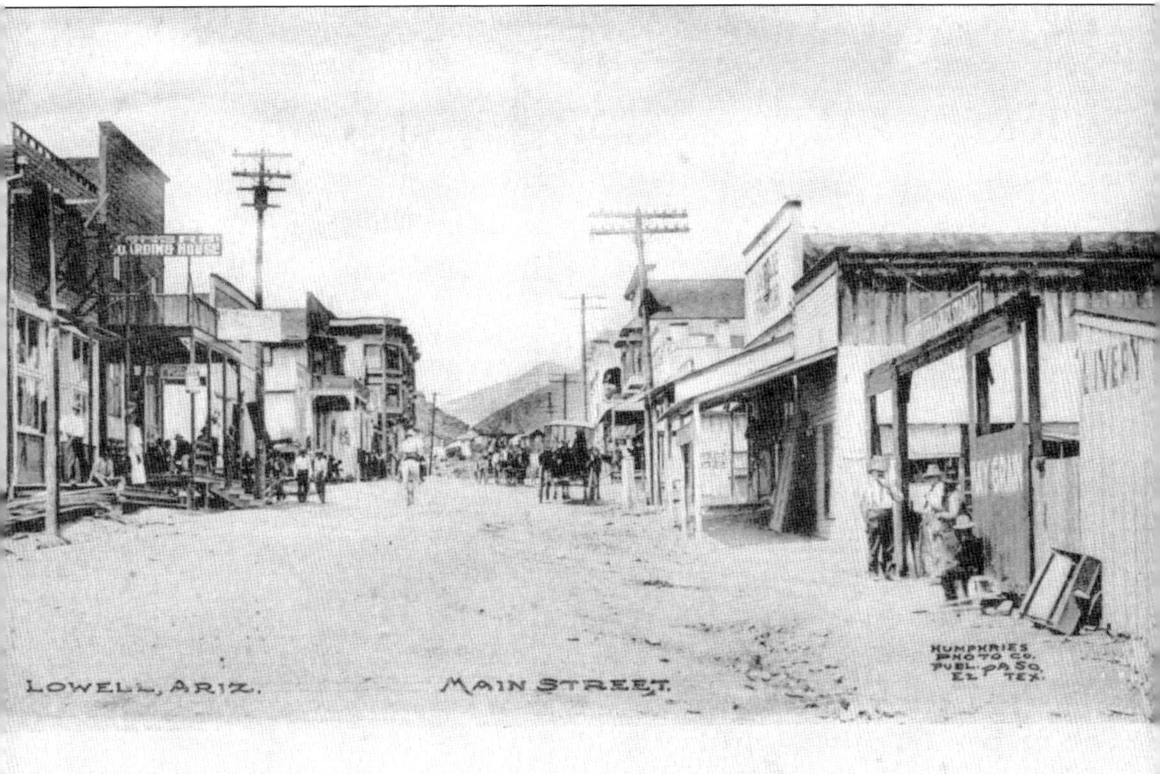

The development of the commercial district in Lowell resembled that of most mining camps, including nearby Bisbee. Erie Street, pictured here in 1902, was home to a livery stable, restaurants, and boardinghouses. Of course, miners could not survive on these things alone, and by 1908, the street offered eight saloons where a miner could quench his thirst after working an exhausting shift. In addition to bars, "houses of the ill-famed" were located in Lowell that provided female companionship for the miners who had needs of a personal nature. However, as the Warren Mining District continued to attract families to the area, these establishments came under attack by members of the local morality movement. A 1904 article in the *Bisbee Daily Review* stated, "Lewd women [would] not become a feature of the rapidly growing suburb," and those who attempted to do such business in the area would be "met with hostility from the start . . . and [would] be arrested on every pretext offered until they are compelled to give up and get out of the community."

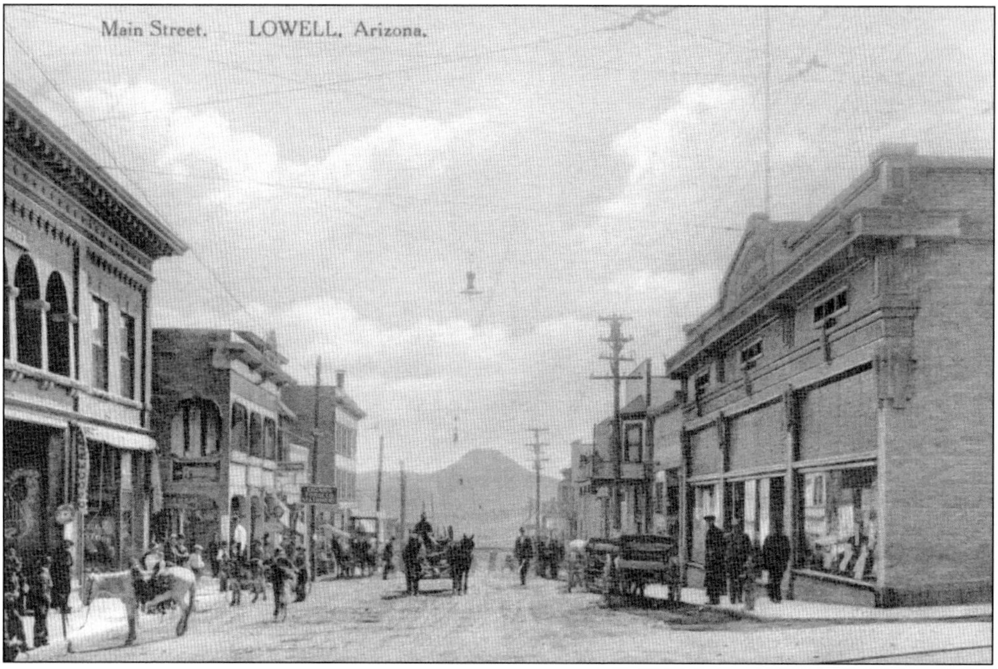

The mines in the Lowell area would go on to employee approximately 2,000 men, and Lowell was noted by the *Bisbee Daily Review* as being the largest and most important of Bisbee's suburbs. However, Lowell was an independent community, and residents did not desire to be an incorporated part of Bisbee. The area was considered a "precinct" and did not have local government in place, only that provided by the county. There was a sensible reason Lowellites did not desire to become an official part of Bisbee: taxes. Lowell had all the amenities of a successful mining camp including schools, a strong commercial district, post office, and volunteer fire department. By being annexed into Bisbee, the residents would not have much to gain except taxation.

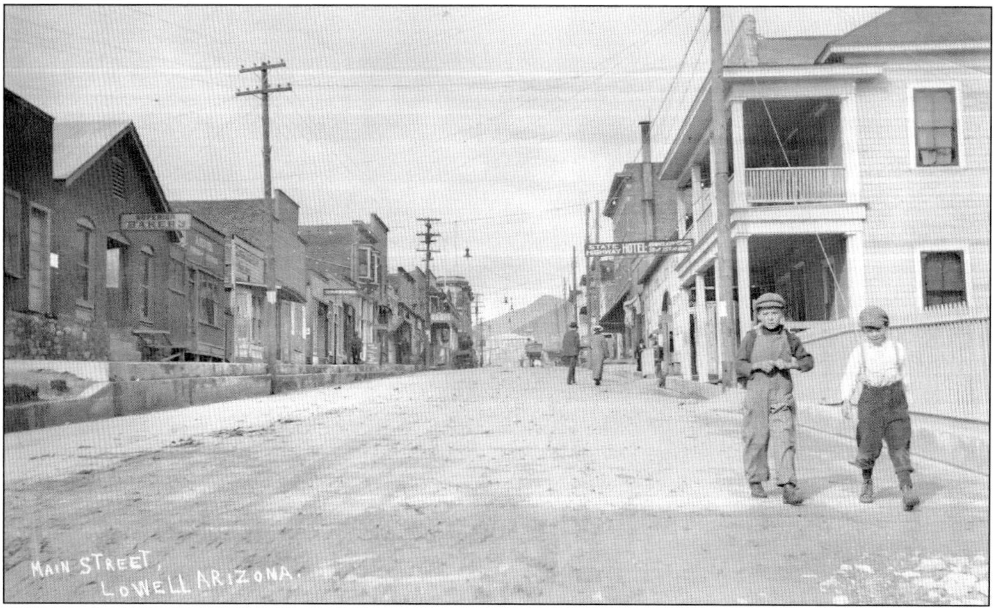

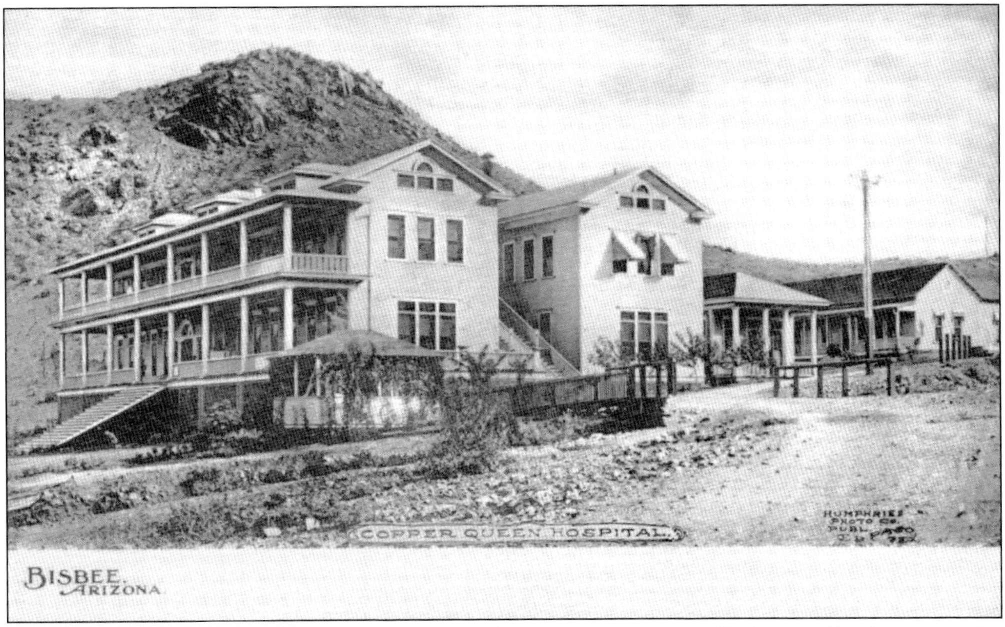

In addition to the mines, neighborhoods, and businesses in Lowell, it was also home to the Copper Queen Hospital. The Copper Queen Mining Company established the district's first hospital in 1900. The structure, shown above around 1902, was located near the western side of Sacramento Hill, but the site was being mined from below, and the ground was sinking, so plans were set into place to physically move the hospital buildings to Lowell in 1909. Each of the institution's structures were lifted from their foundations, loaded onto railroad cars, and transported about one mile to Lowell. The hospital, as seen below in the 1910s, remained in Lowell until 1929, when it moved to a new facility in Bisbee because mining caused the ground to sink yet again.

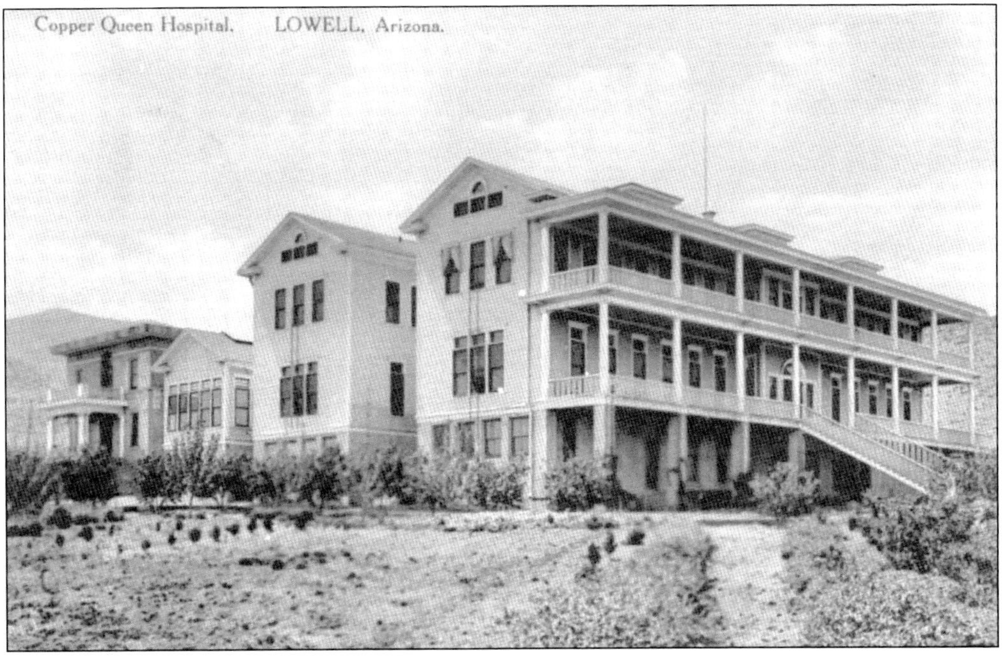

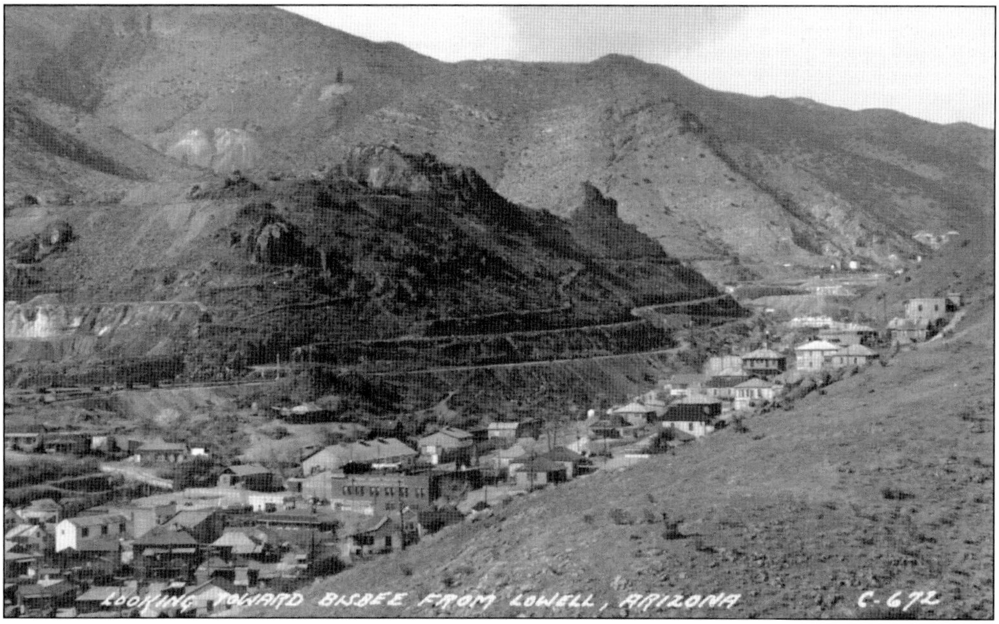

Much like Bisbee, Lowell was described as having an "irregular shape" largely due to its sporadic growth around the area's mine sites. By 1915, the *Bisbee Daily Review* reported that Lowell maintained a population of between 5,000 and 6,000 inhabitants, which included the neighborhoods of Johnson Addition, Upper Lowell, and Jiggerville. Johnson Addition, pictured in the above 1930s card, was northwest of Lowell and was developed by real estate investors to accommodate the housing needs of the ever-growing Warren Mining District. To the south of Johnson Addition was the neighborhood of Upper Lowell, shown below around 1905, which was within close proximity to both the area's mines and commercial district.

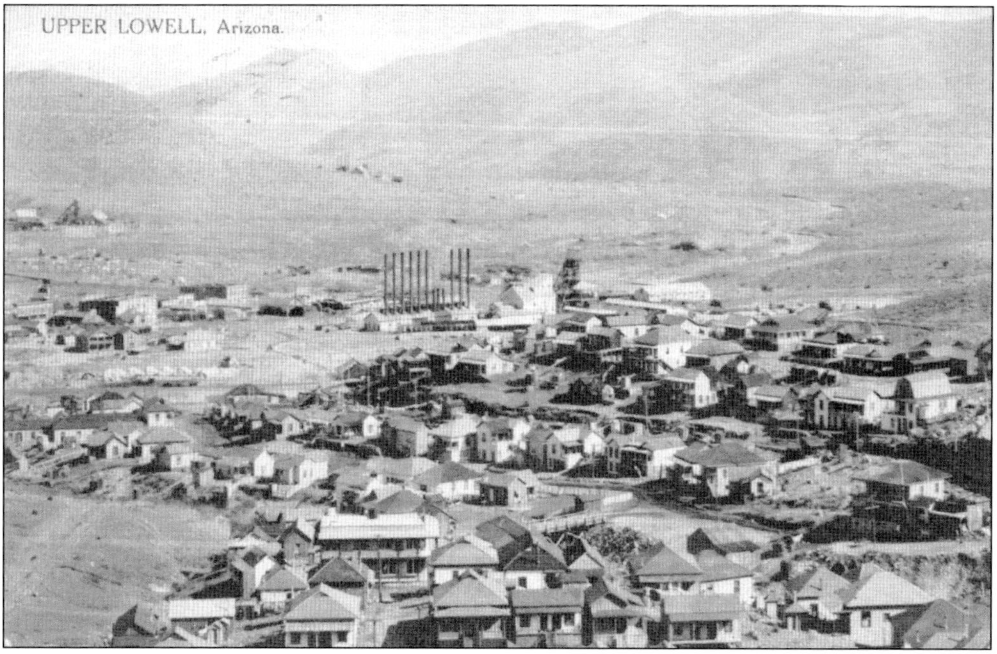

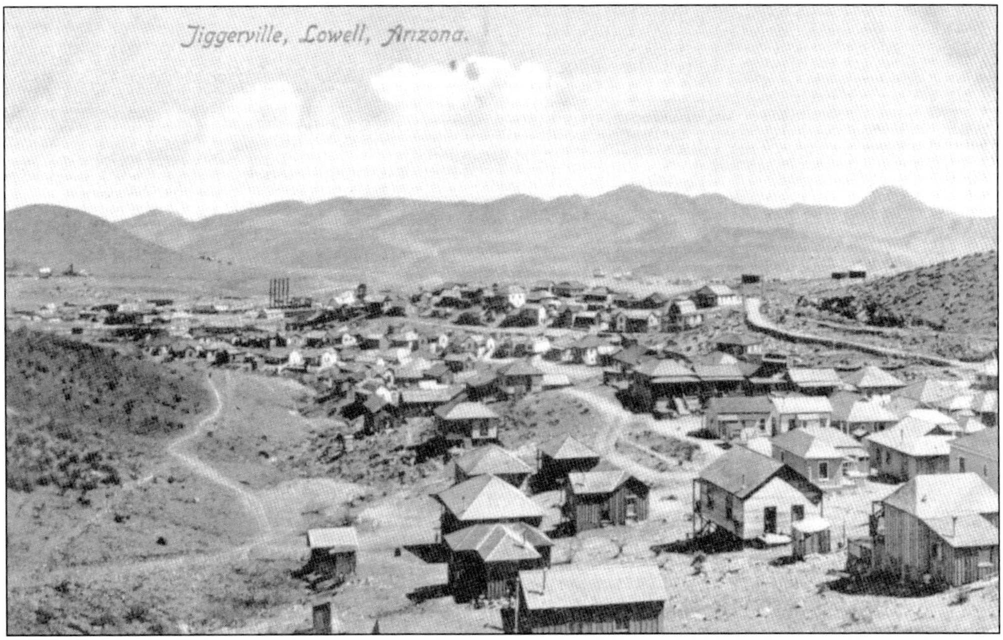

Heading southwest from Upper Lowell was the neighborhood of Jiggerville. The name of the community derived from the word Jigger, a term used for shift bosses who worked for the mining companies. Because the area was a very short distance from the mines, many bosses lived in this area. Living on the edge of active mine sites, the reason for the community's existence was never far from the minds of residents. People recalled hearing the noise of the machines on the surface night and day, as well as having the occasional boulder roll into the neighborhood from the nearby mine waste piles. Residents of Jiggerville understood the community would one day disappear, as the mining companies' quest for ore included the land their homes sat upon.

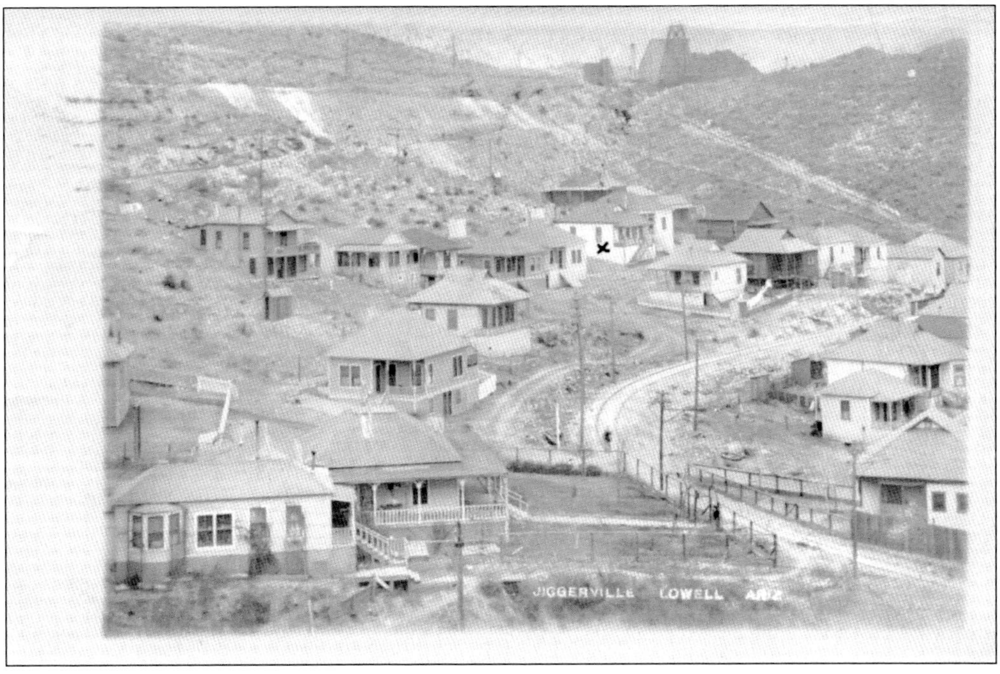

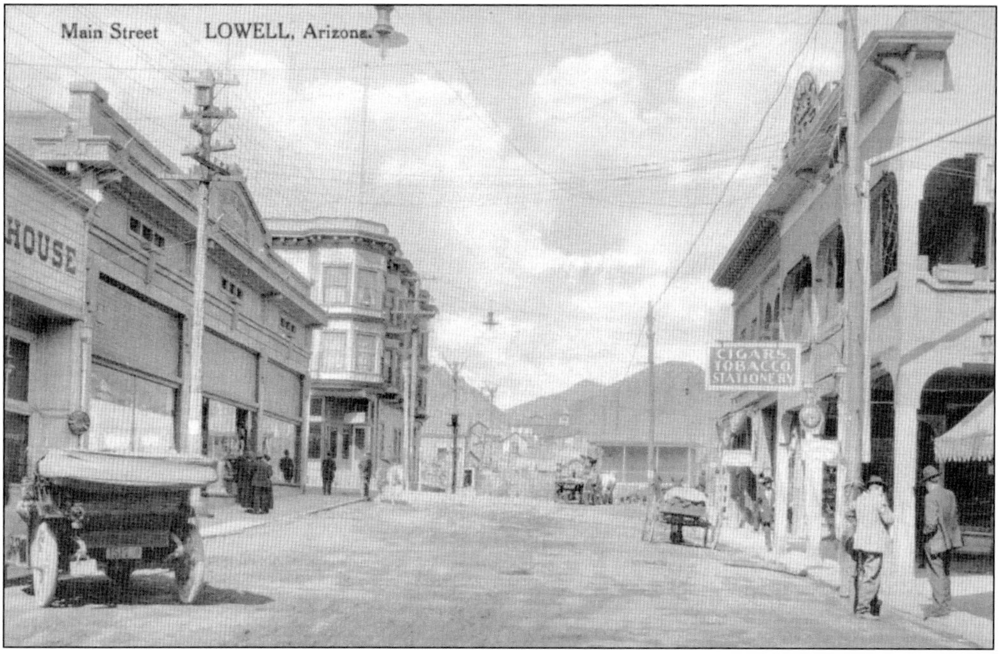

Despite knowing the neighborhoods in Lowell had an impending expiration date, people continued to move to the area, and businesses continued to be established in the precinct. The community was described in 1917 by the *Bisbee Daily Review* as "one of the most energetic towns in the state and [had] a payroll that would dazzle many cities a great deal larger." Responding to the prosperity of the area, businesses continued to emerge in Lowell, including a branch of the Phelps Dodge Mercantile, shown on the left in the above 1910s image. Another well-known business in Lowell was the Bisbee Lumber Company, with its lumber yard visible in the bottom left of the 1930s postcard below.

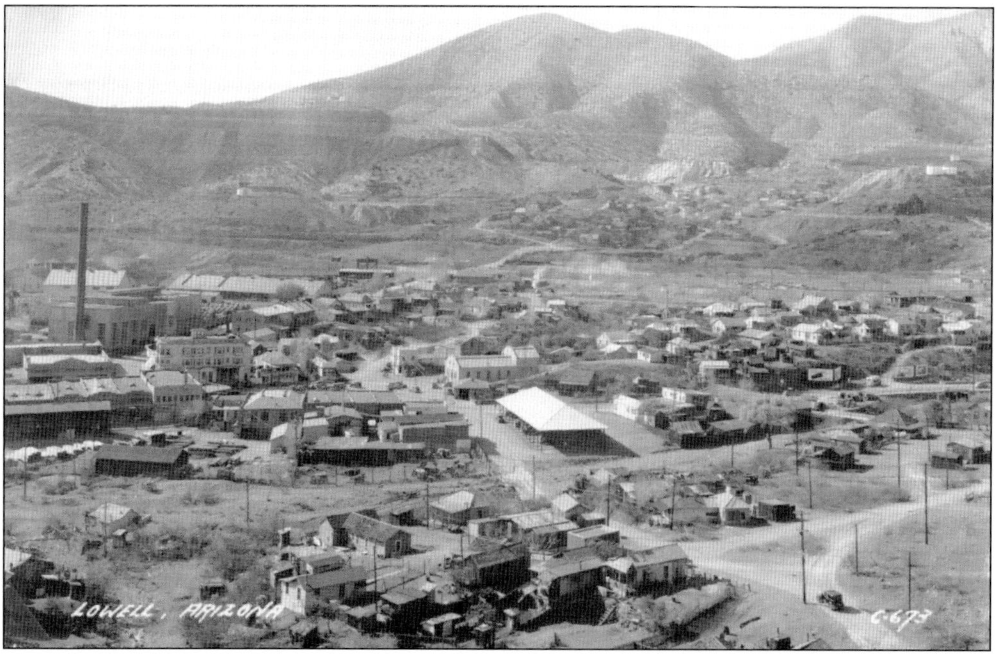

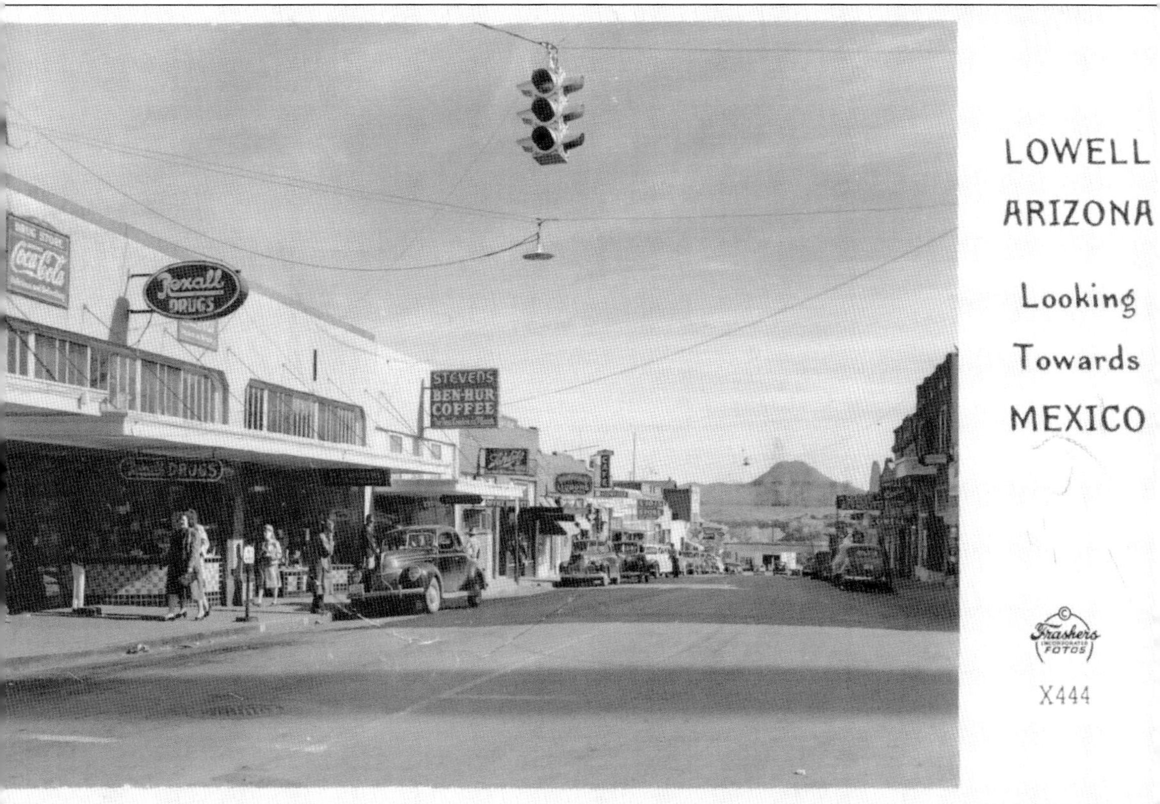

In a 1917 edition of the *Bisbee Daily Review*, the newspaper stated that Lowell was a "lasting town built upon a firm foundation, the copper industry." However, the source of the community's success would also be its demise. Despite being a vibrant area that was home to thousands of miners and their families, the day eventually came in the 1950s when the company announced plans for a new surface mining project. The undertaking would become the Lavender Pit and involved developing the areas occupied by Lowell, Upper Lowell, Johnson Addition, and Jiggerville. Residents were offered the option of either receiving the market value for their home, which would then be demolished, or the company would pay to have the structures moved to a new site. While Johnson Addition, Upper Lowell, and Jiggerville vanished with the Lavender Pit, a fragment of Lowell's commercial district, Erie Street, still stands. It is pictured here in the 1930s.

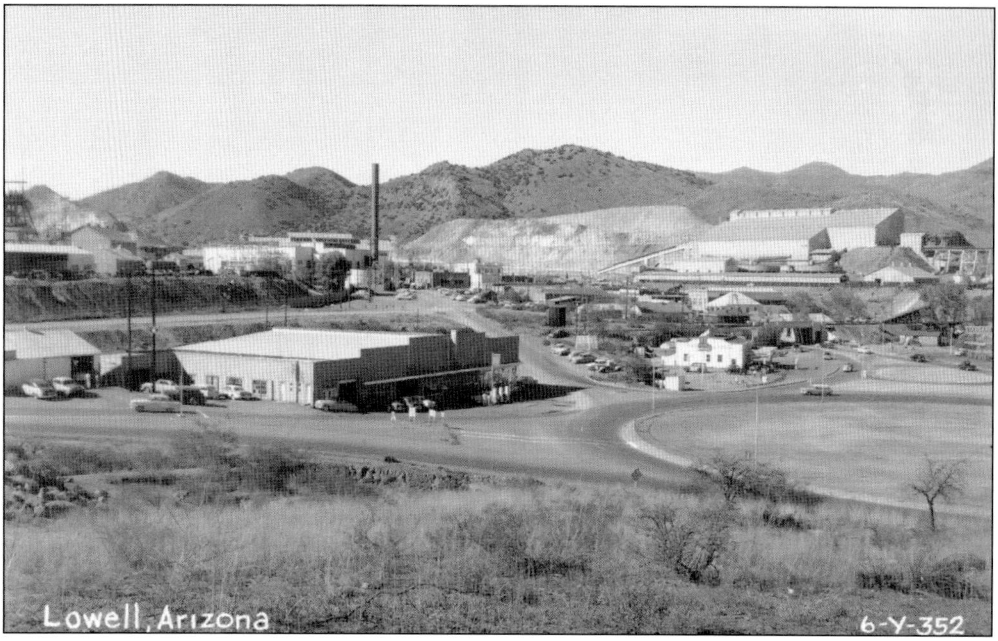

Other portions of the community that would survive the creation of the Lavender Pit included the traffic circle visible above in the 1950s. The traffic circle was and continues to be busy with automobile traffic, as it is the junction where the roads to Douglas, Lowell, Bisbee, and the San Jose area converge. A section of Lowell to the east also remained intact, seen in the right of the 1930s postcard below. This area was an important part of the community, as it was the location of Lowell School and Evergreen Cemetery.

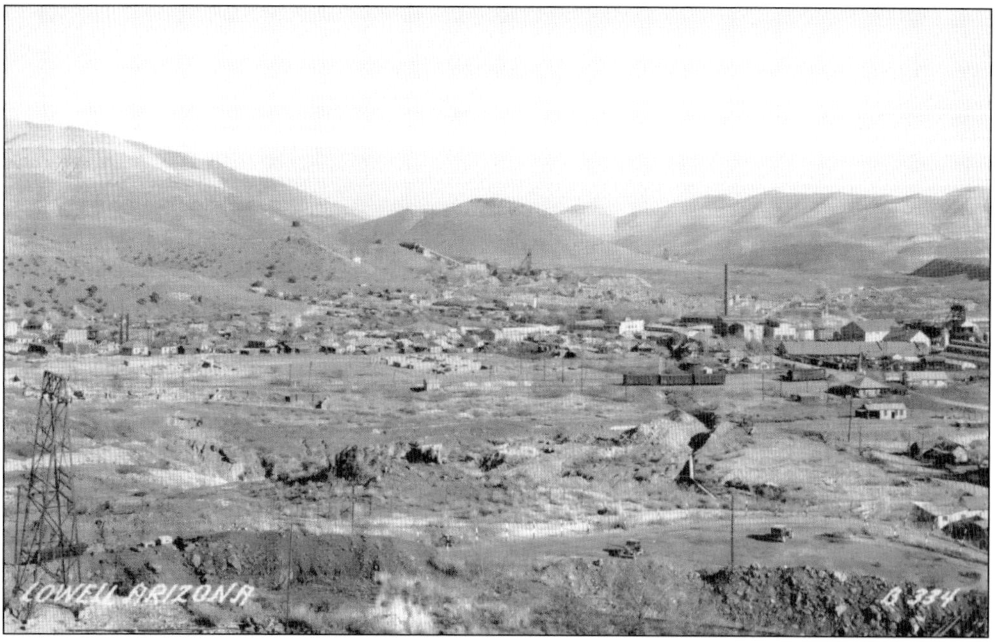

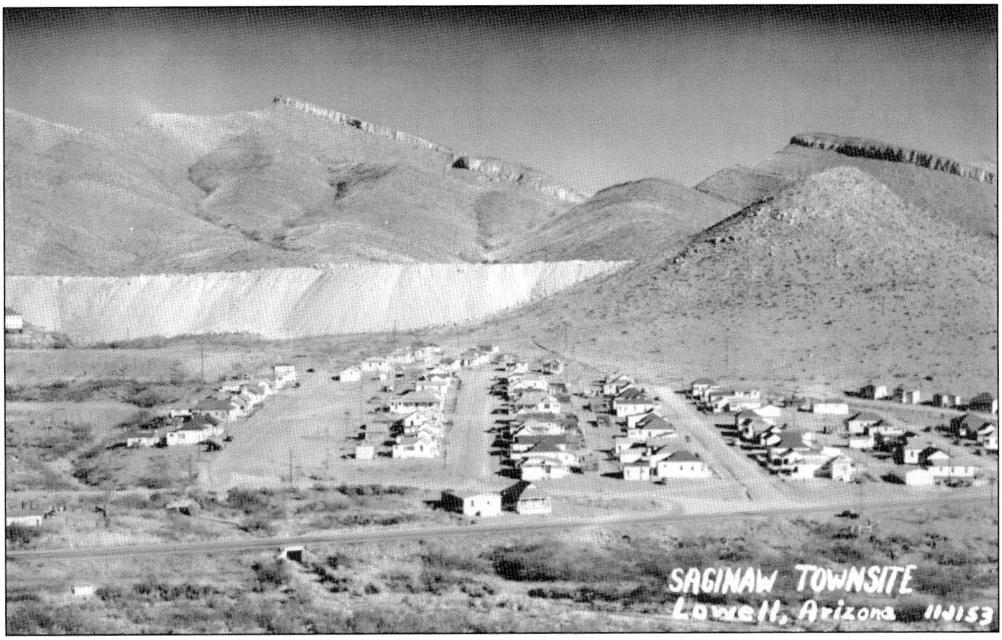

Most of the homes that were moved from the Lowell region were relocated to Saginaw, Galena, and Bakerville. Pictured above in the 1950s with Highway 80 running alongside it is Saginaw. The community is located on the eastern end of the remnants of Lowell. While Bakerville and Galena are a short distance from one another, the Campbell Mine lies between them. Galena is shown in the upper portion of the 1950s postcard below near the communities of Briggs and South Bisbee. Bakerville is visible in the foreground and is located next to Warren.

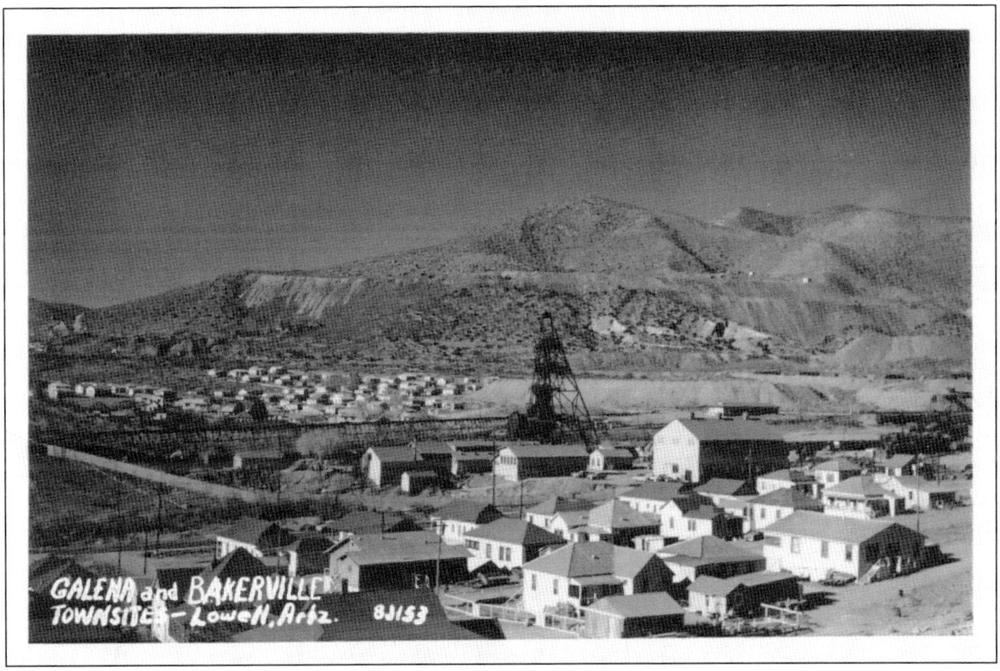

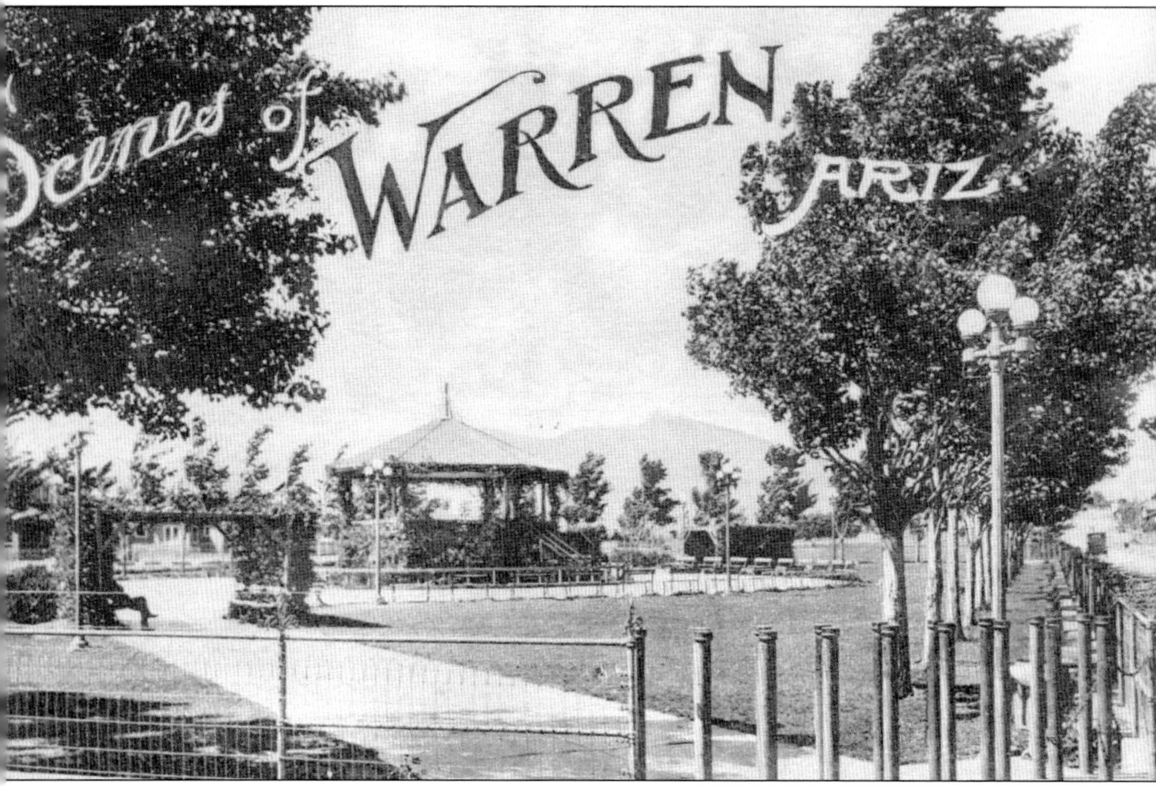

The creation of the community of Warren was unlike any other in the mining district. Most neighborhoods appeared haphazardly around mine sites, resulting in unusually shaped communities located in areas with problematic terrain and often poor sanitary conditions. However, the new townsite was different from these neighborhoods. Warren was the vision of the Calumet & Arizona Mining Company, which stated in its 1906 annual report that, in order to attract and retain "the best class of employees, a desirable place be provided where they can build and own homes." To achieve this goal the Calumet & Arizona established the Warren Company, which oversaw the creation of a new planned community offering modern homes and recreational amenities. Moreover, to guarantee the desirability of living in an area a considerable distance from many of the district's mines, a streetcar service was established to transport residents in a quick and inexpensive manner.

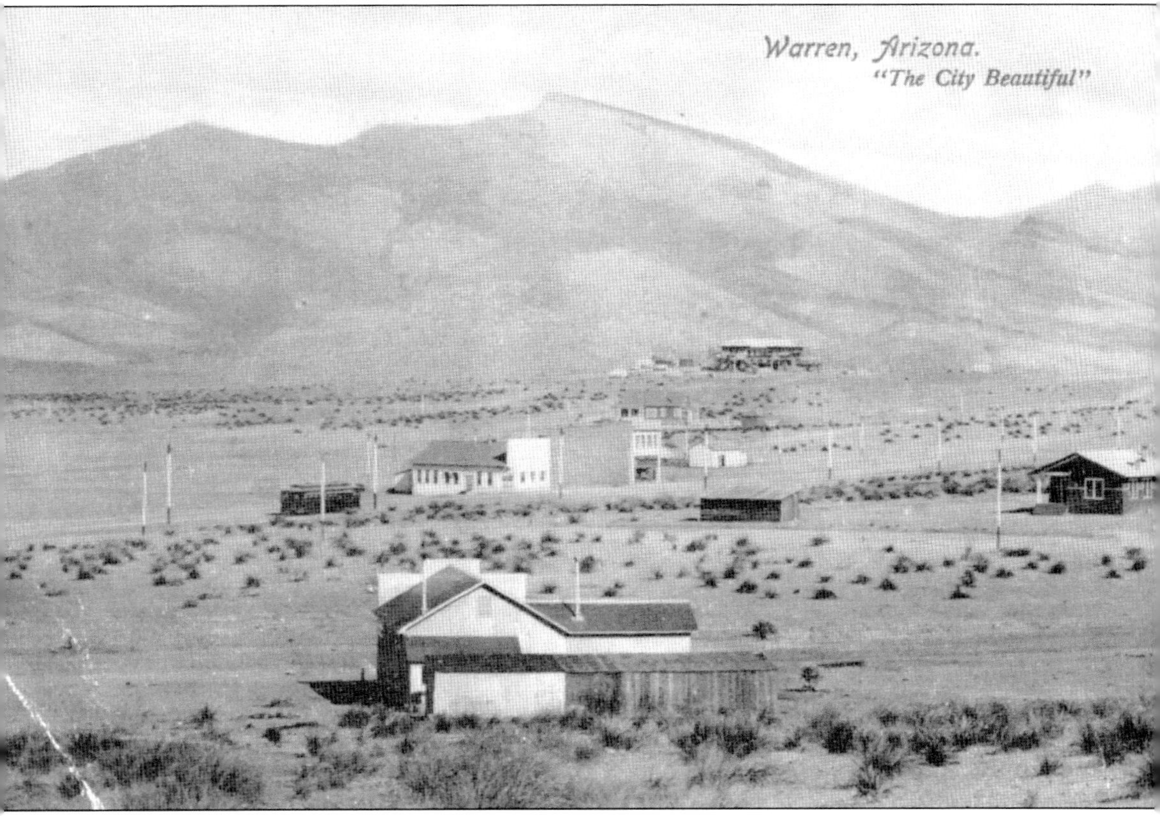

To produce the most attractive community possible, the Warren Company called upon the services of Warren Henry Manning, a landscape architect and city planner. Manning was a proponent of the City Beautiful movement during the 1900s in the United States. The concept entailed creating communities that were not only clean, orderly, and aesthetically pleasing, but that also enhanced the overall moral and civic nature of a city. When Manning arrived in Bisbee, he wrote his wife and declared he gazed upon "such a town you never saw in your life." Explaining that Bisbee was "all on edge except where the edges meet at the bottom" with houses "arranged in . . . every possible angle as though they had been tumbled down the hillsides and then patched up and made into a house where they happened to land, [and] streets that wiggled through where they could be squeezed in." The proposed plan Manning put forth for Warren guaranteed the community would not share such disorderly characteristics with Bisbee.

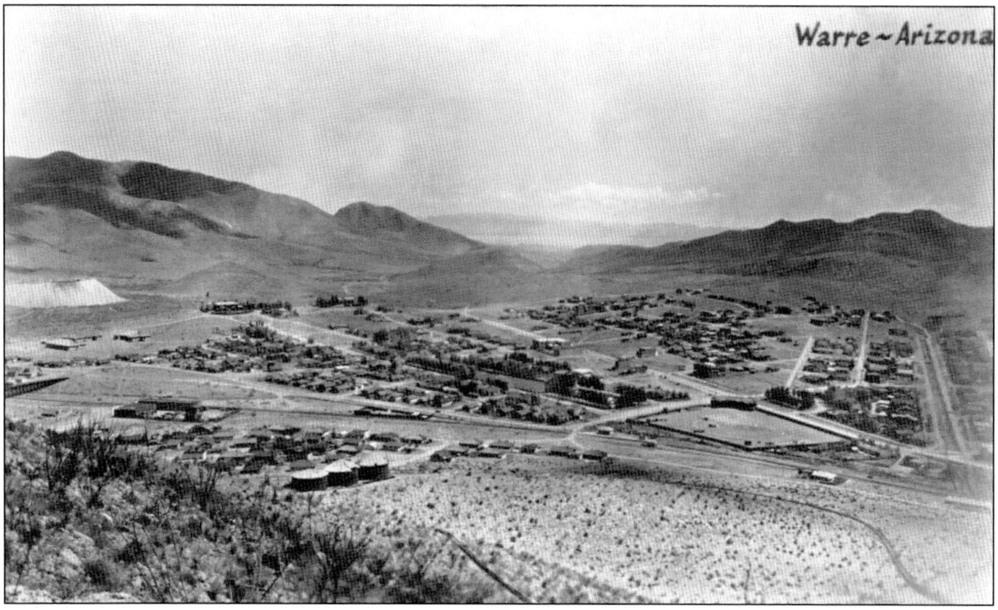

Manning designed Warren in a fan shape, in which the city would radiate out from a common point. Looking from the lower right corner of the 1920s postcard above, this design is clearly visible. The community offered city improvements such as maintained roads, public transportation, and homes that included modern sewer systems, electricity, piped-in water, and city fire protection. While these are basic features in towns today, these aspects were not prevalent in neighborhoods or homes in Bisbee during the 1900s. On October 6, 1906, Warren's ground breaking took place, with Bakerville's oldest resident, Mrs. S.E. Hazzard, removing the first shovelful of dirt. Warren would no longer be a barren desert landscape, but a new, modern community for the district's residents. The image below captures Warren's growth in the 1910s.

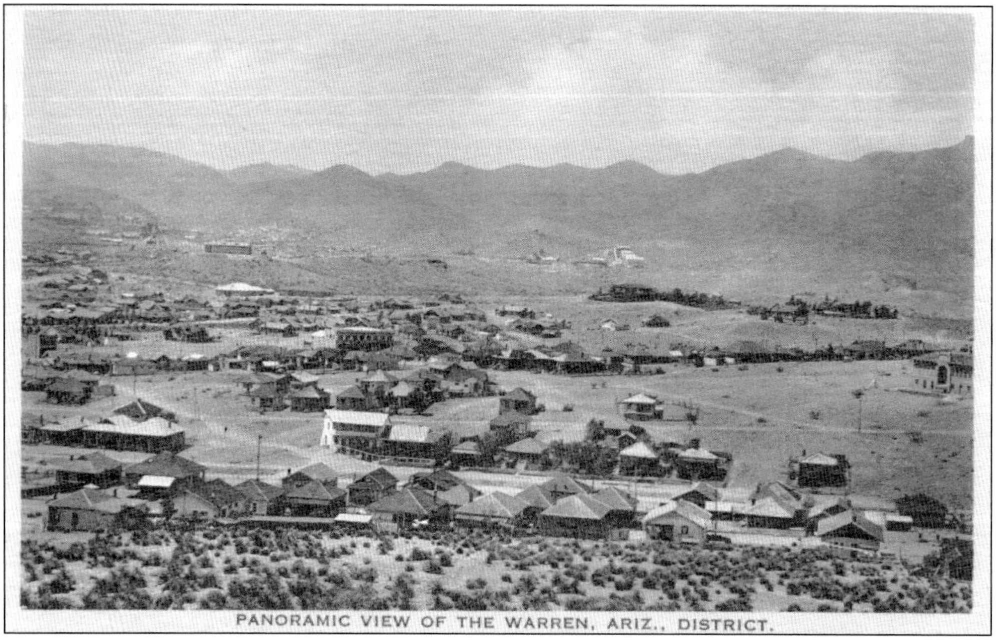

PANORAMIC VIEW OF THE WARREN, ARIZ., DISTRICT.

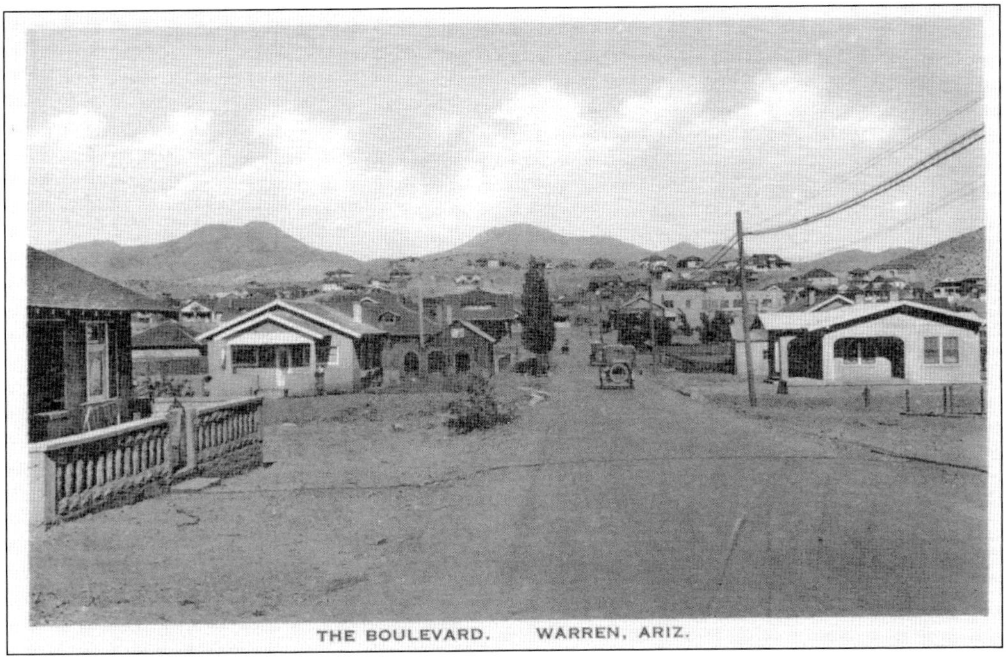

The townsite was named after George Warren, the prospector of questionable principles, who is credited as one of the discoverers of the mining district. However, unlike its namesake, Warren was marketed as a place where a family would be surrounded by a decent and proper environment. An advertisement in a 1907 issue of the *Bisbee Daily Review* asked readers, "Do moral influences surround your home? If not, why not?" Warren was promoted as a community that offered individuals the "Opportunity to give [their] children a proper environment." The postcard above captures Warren in the 1910s, while the image below was taken in the 1930s.

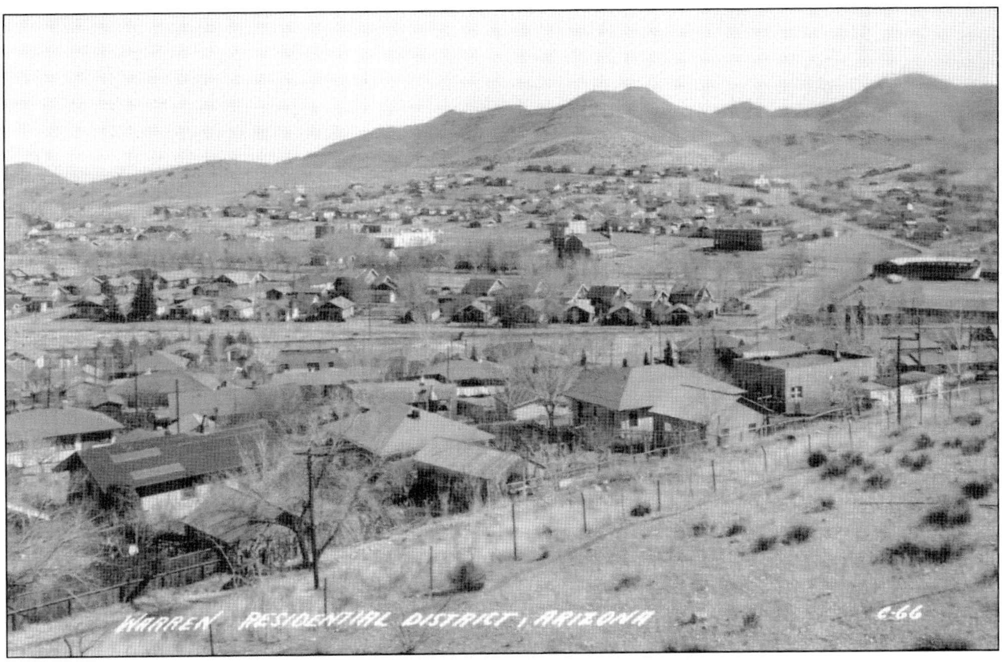

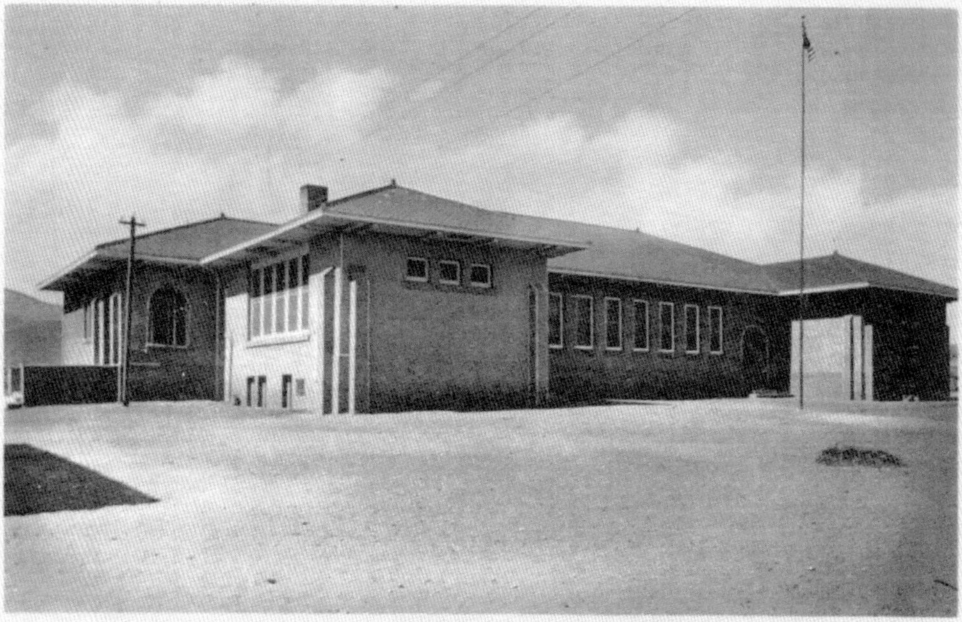

GREENWAY SCHOOL. WARREN, ARIZ.

Schools and parks were thought of as vital components to living a moral lifestyle, and Warren had an abundance of these features for residents. Greenway School, pictured above around 1920, is perhaps the best-known school in the community and continues to operate today. Greenway School opened in 1917 and was named after John C. Greenway, the general manager of the Calumet & Arizona Mining Company, as well as a resident of Warren. Outdoor spaces were also believed by the planners of Warren to add to residents' quality of life, and Warren's Vista Park provided ample space for relaxation or recreation. A portion of the park is visible in the center of the 1912 postcard below.

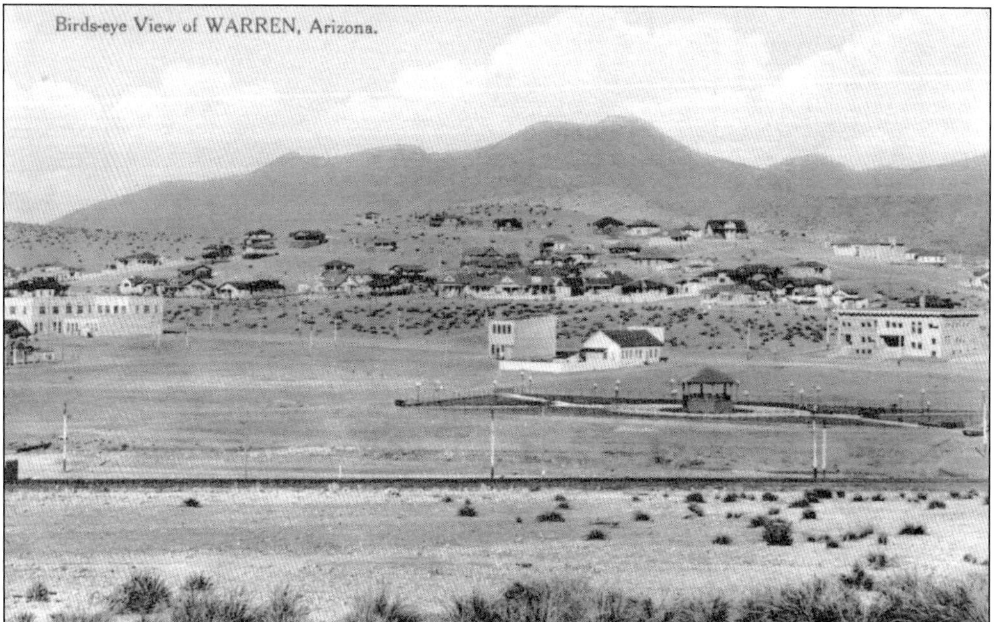

Birds-eye View of WARREN, Arizona.

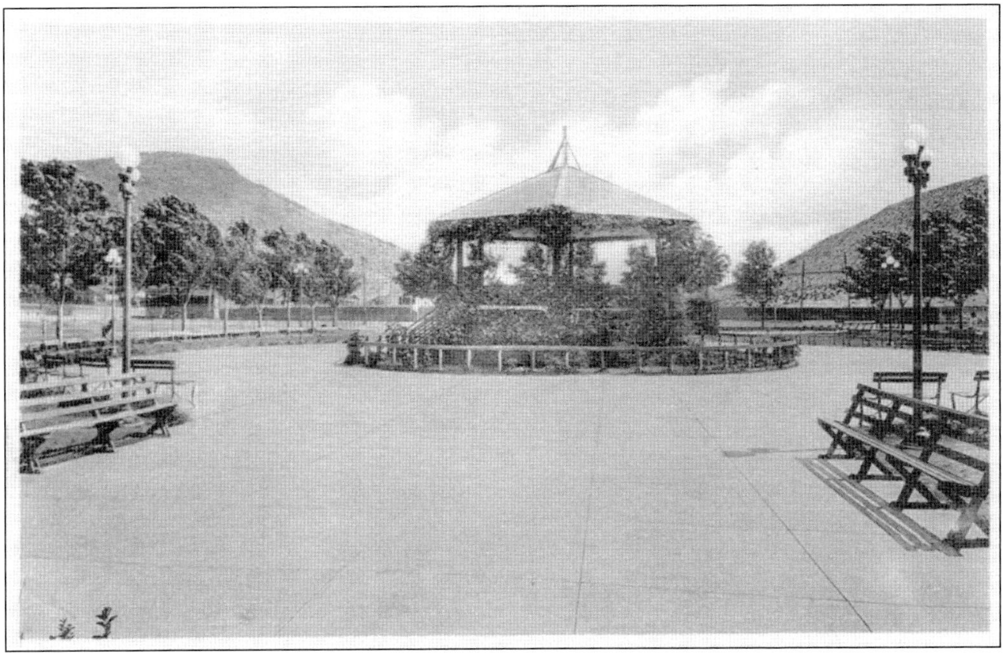

The design of Vista Park was inspired by the McMillan Plan, used to design the layout of Washington, DC. Using the outdoor space as the centerpiece of the community, the park would ultimately become an impressive 2,500 feet long, 160 feet wide, and flanked by two boulevards. The area became frequently used for family and social gatherings. The southern end of the park, shown above in the late 1910s, was often the site of dances and celebrations. One of the best attended events at Vista Park was the Labor Day celebration of 1917, shown below. The *Bisbee Daily Review* reported that 10,000 people made their way to Warren that day to join in the festivities.

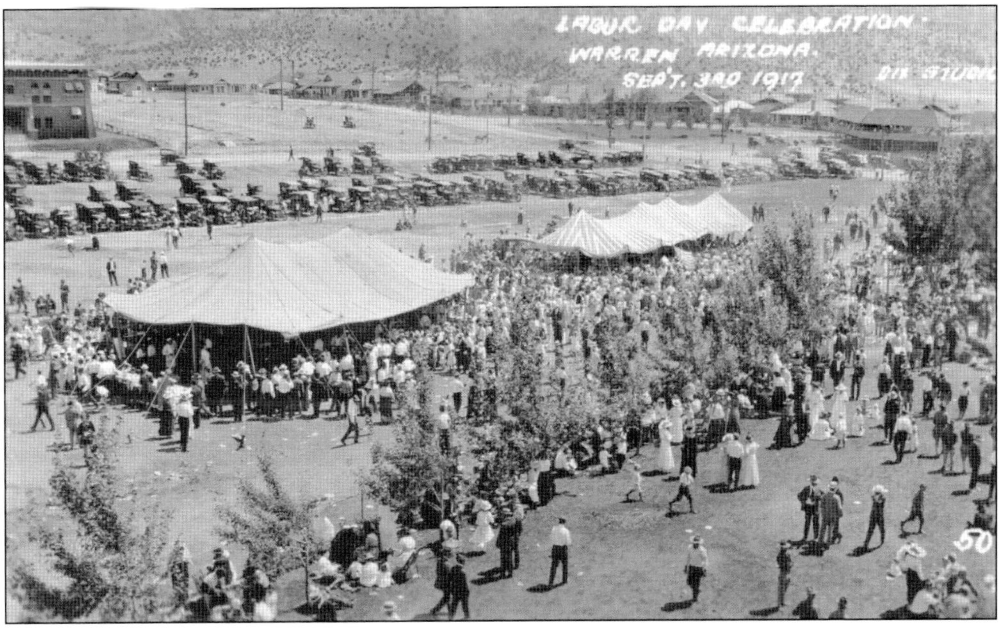

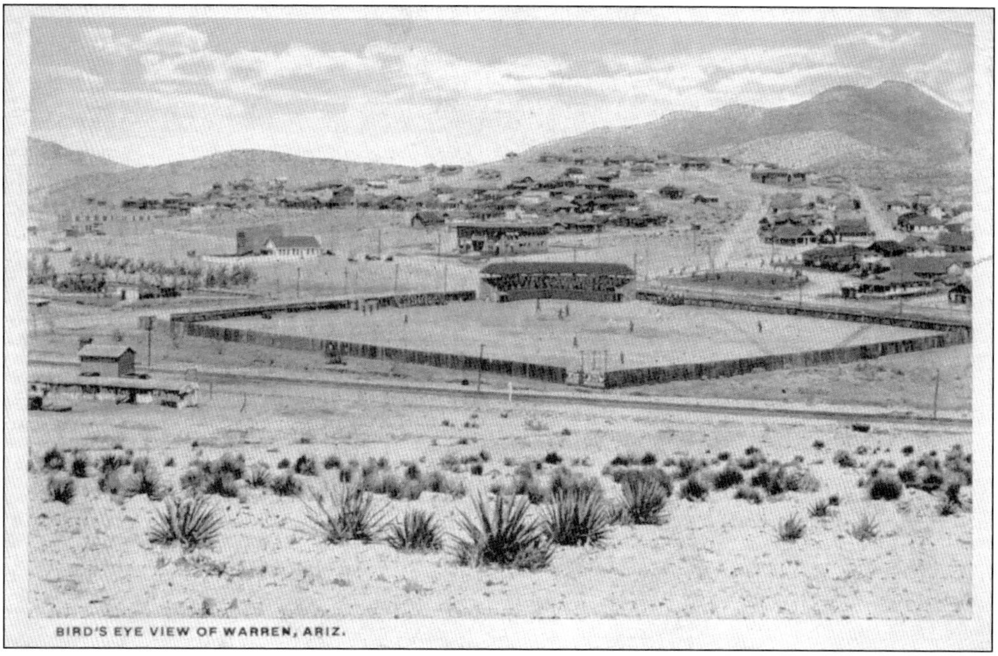

The south end of Vista Park was home to another recreational amenity, the Warren Ballpark. Baseball was a beloved pastime of the district's residents, and a ballpark was included as part of Warren's design plan. The first game at the ballpark took place in 1909; it went on to be used by schools, mining companies, and city teams, and is the oldest ballpark in Arizona. The Warren Ballpark was also the location where strikers and union sympathizers were taken during the Bisbee Deportation in 1917. Being located next to the railroad tracks, the site was chosen for sheriff's posse members to assemble the strikers. The individuals who agreed to return to work at the mines were released; those who refused were forcibly loaded on a train and removed from the area.

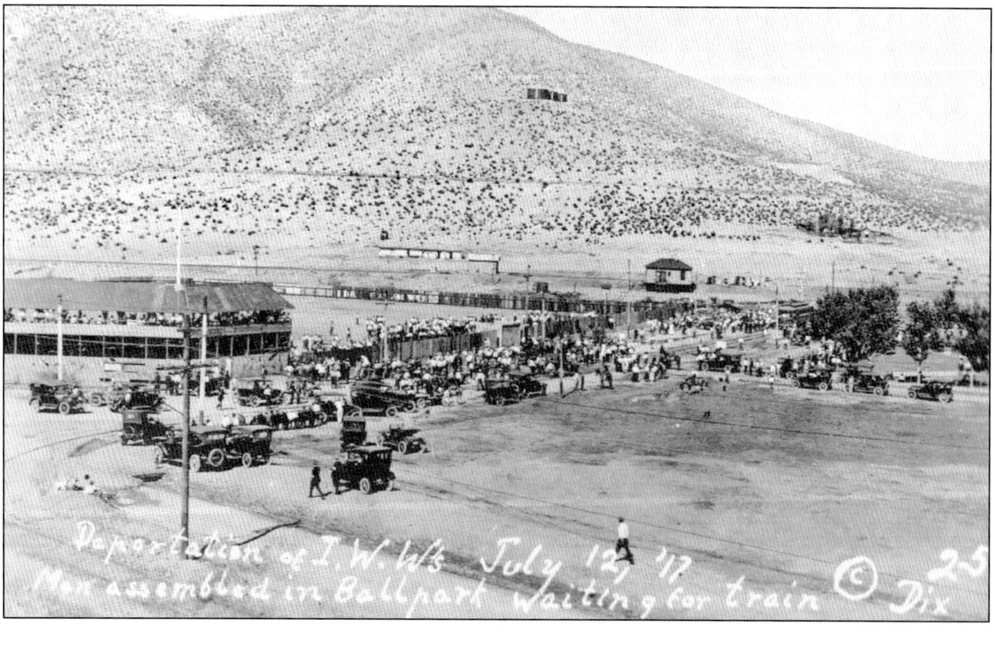

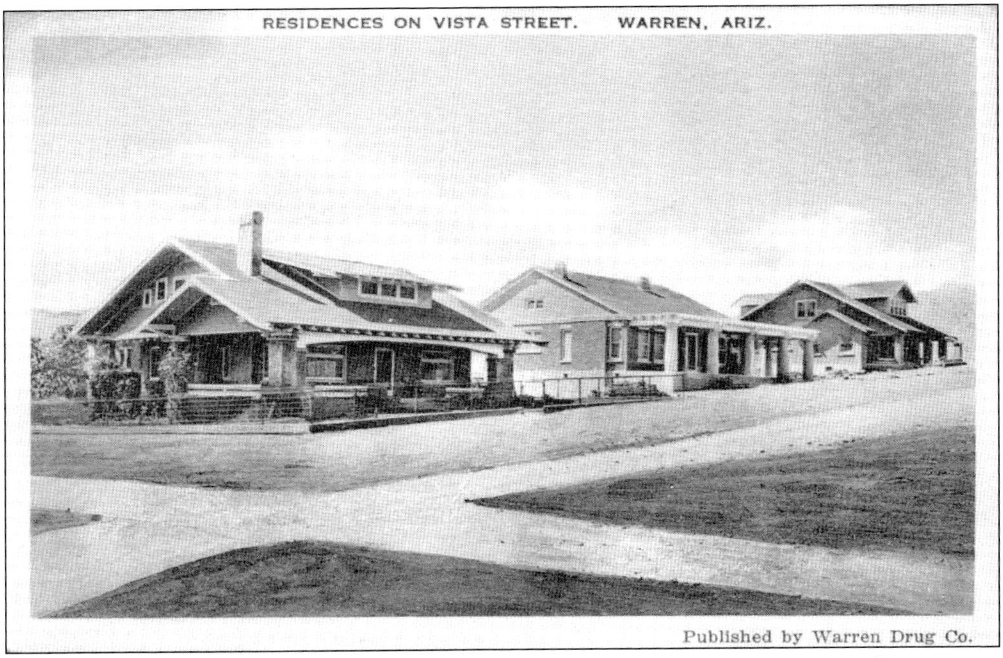

In 1909, the *Bisbee Daily Review* reported that since building began in Warren, new homes were going up at a rate of one per week and that it was "one of the prettiest towns in the southwest assured steady and healthy growth." It grew indeed, and by the time these 1910s images were captured, Warren was home to several hundred residents. The civic improvements, parks, and beautiful homes were attractive incentives to move to the area. The bungalow was a popular architectural style used in the townsite, and some of the finest examples of these types of homes lined the boulevards along Vista Park.

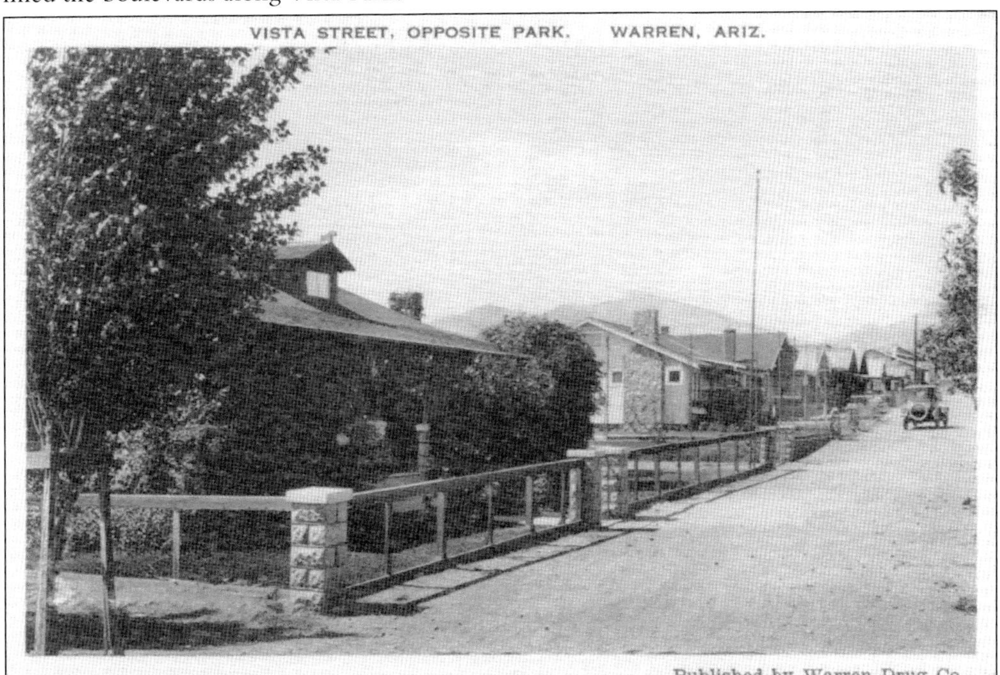

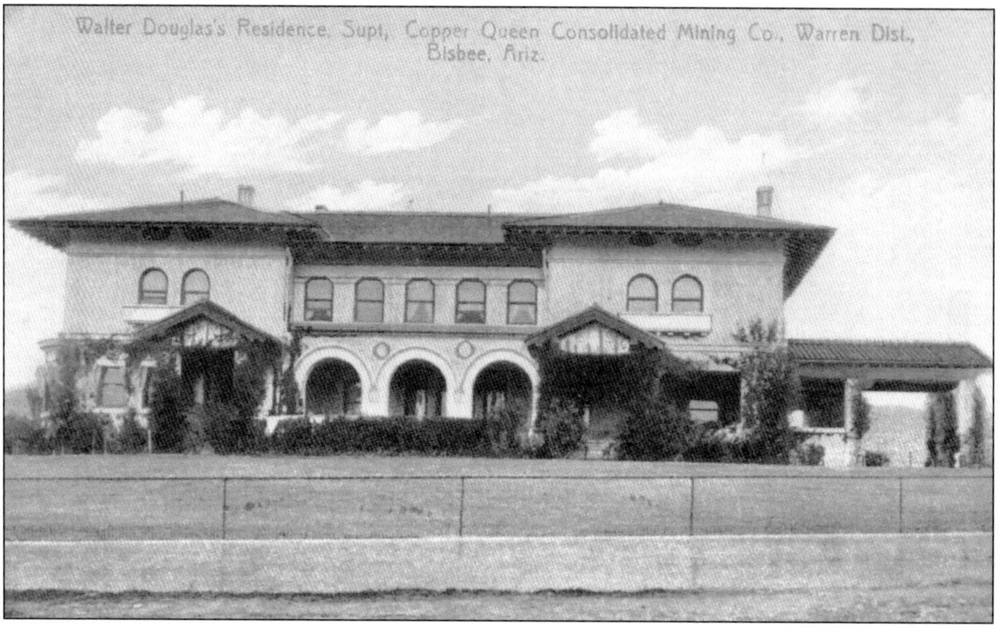

Warren was also viewed as a desirable place to reside by members of the mining companies' management, and many moved to the community. This included the most powerful man in Bisbee, Walter Douglas, the manager of the Copper Queen Consolidated Mining Company. The indisputable crown jewel of Warren, the Douglas residence was erected at the north end of Vista Park and offered expansive views to the south of not only the park but also the mountains in Sonora, Mexico. Construction was completed on the impressive two-story, 41 room, Mission Revival–style home in 1908 at a cost reported to be over $65,000. However, the end product resulted in what the *Bisbee Daily Review* called the handsomest and most substantial home in the territory.

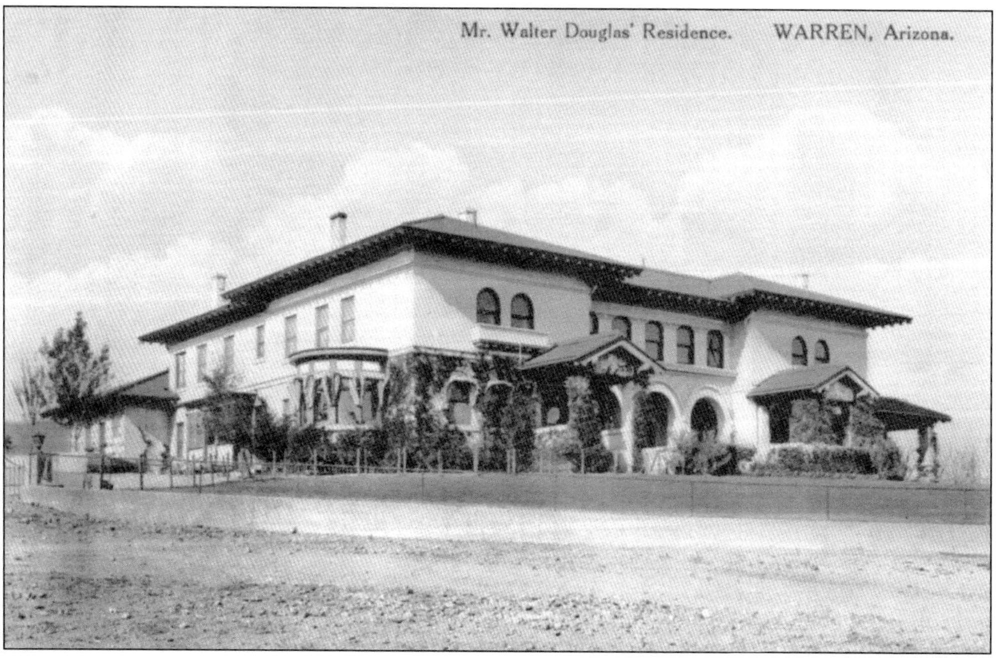

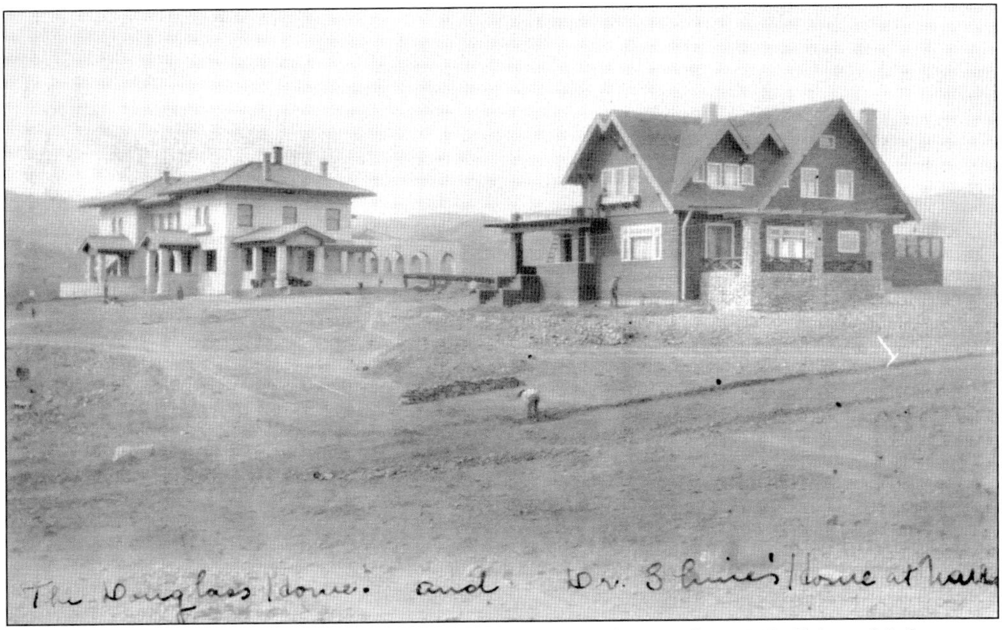

Another notable home constructed in Warren was the residence of Dr. F.E. Shine, chief surgeon of the Copper Queen Consolidated Mining Company. Dr. Shine's large California Bungalow–style home was located next to the Douglas residence and is visible on the right in the above 1908 image. Another member of the district's mining management who called Warren home was Louis Powell. Powell was the manager of the Calumet & Arizona Mining Company when the townsite was developed. However, the California bungalow is best known as being home to another resident, Col. John C. Greenway, Powell's successor as manager of the Calumet & Arizona. Greenway served in the Spanish-American War as one of Theodore Roosevelt's Rough Riders and went on to build an impressive mining career in Arizona.

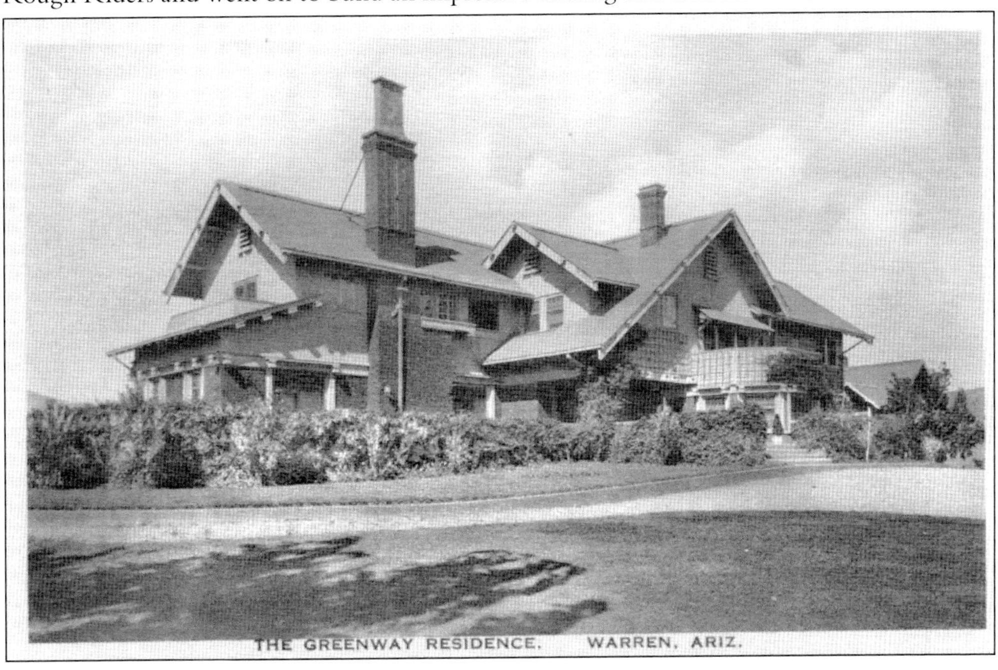

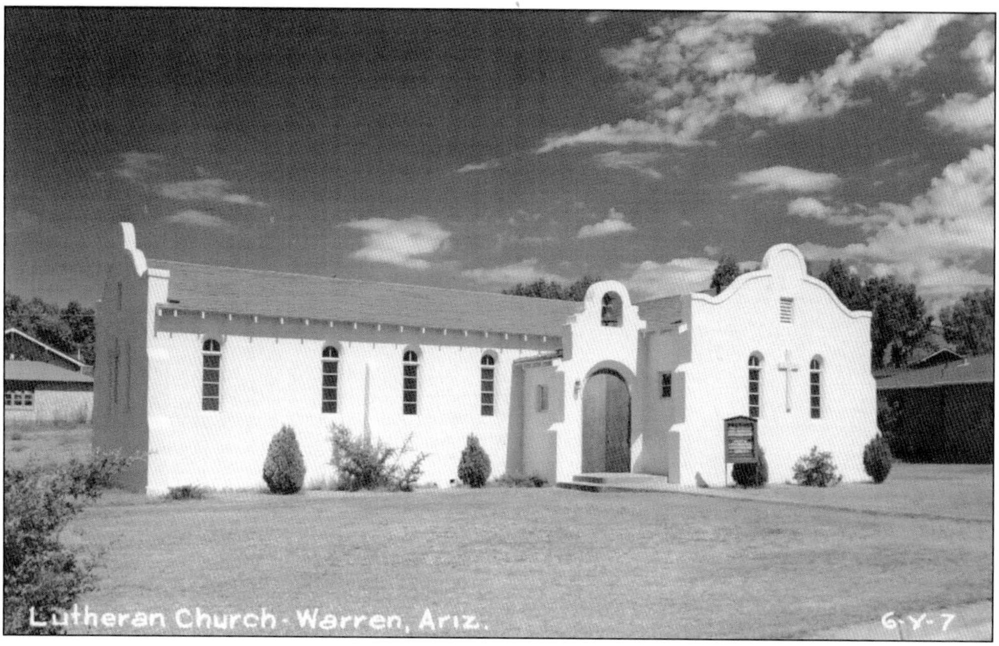

Due to the diverse population in the mining district, numerous religious denominations had houses of worship in the area. Keeping in line with providing a "moral" community to raise one's children, many of these churches were established in Warren. This included Grace Lutheran Church on Arizona Street, pictured above in the 1940s. The Community Church of Warren, located to the south of Grace Lutheran on Arizona Street, opened its doors in 1925 to Christians of all denominations.

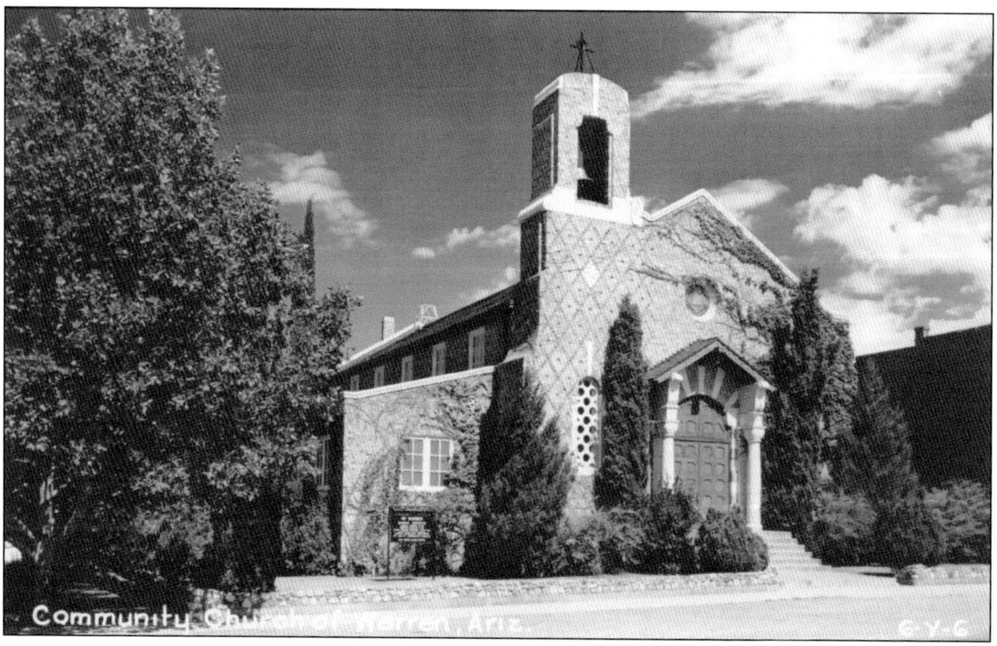

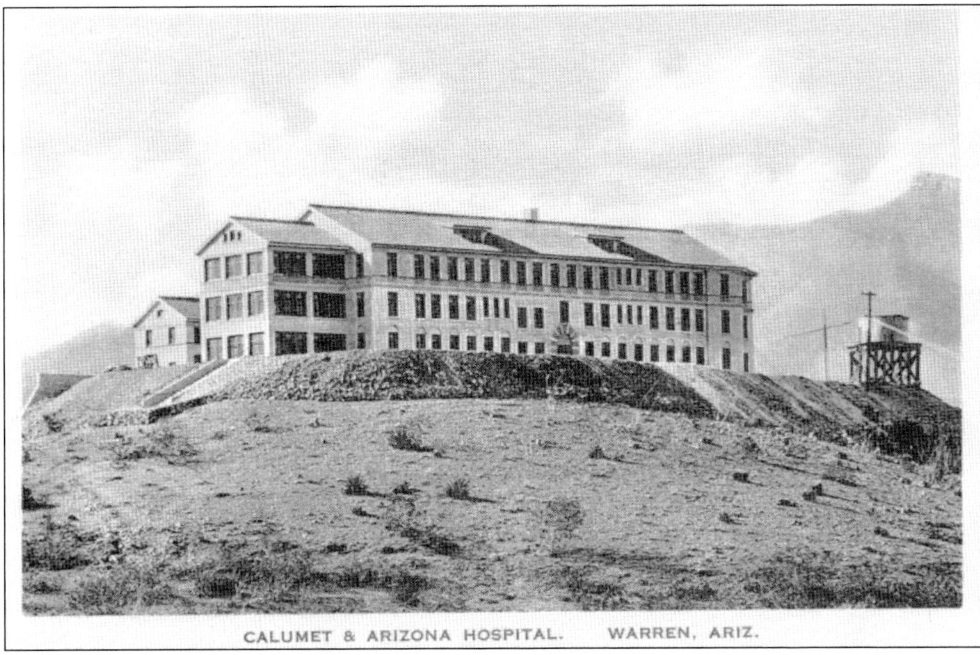

On the top of a hill between Bakerville and Warren during 1918, the Calumet & Arizona built a hospital to care for miners and residents of the district. The *Bisbee Daily Review* thought so highly of the health care facility that it proclaimed, "If there is such a thing as compensation, for injury, sickness or pain on this earth it is here that it can be found."

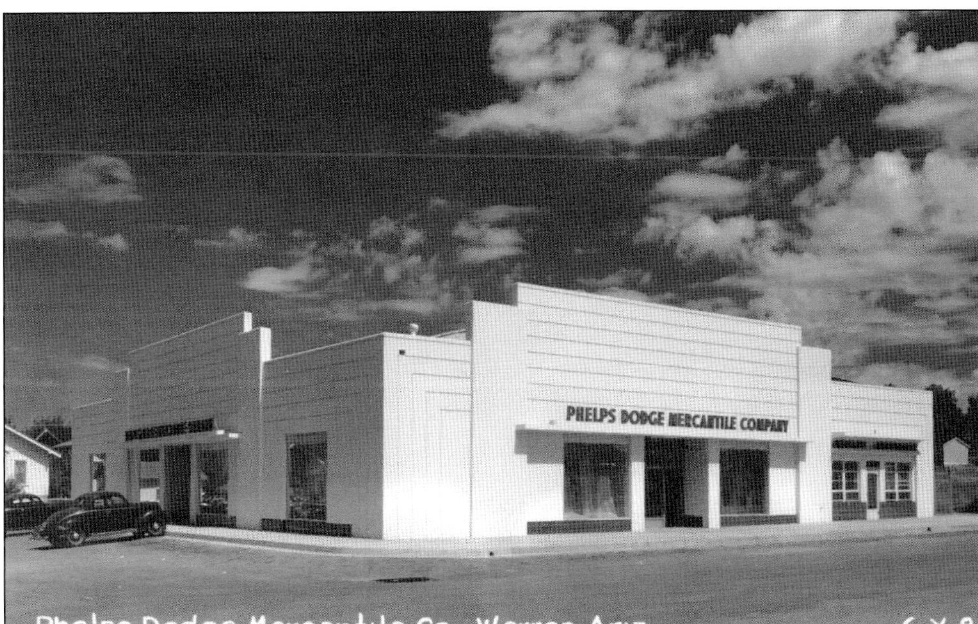

Realizing the revenue potential of the growing residential area, businesses soon followed the residents to Warren. One of course was the mining company, which established a branch of its mercantile store on Arizona Street. Pictured is the Phelps Dodge Mercantile during the 1930s, which has the distinction of being the only Art Deco structure in Warren.

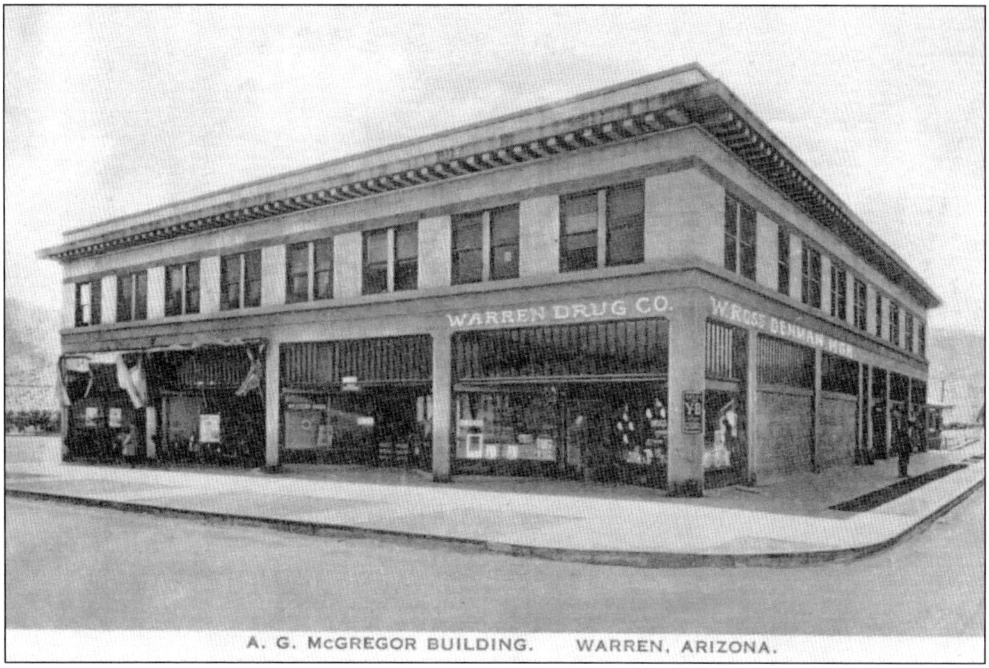

Before it was located in the Art Deco building, the mining company's mercantile store was in the building next door. Over the years, numerous businesses occupied the structure, including a barbershop, located in the center. While the name of the establishment has changed over the years, a barbershop is still in the building many decades later. However, the edifice is perhaps best known for being the home of the Warren Drugstore. Located in the right side of the building, it was the classic pharmacy with a soda fountain patronized by countless children who enjoyed milk shakes and refreshing phosphate sodas.

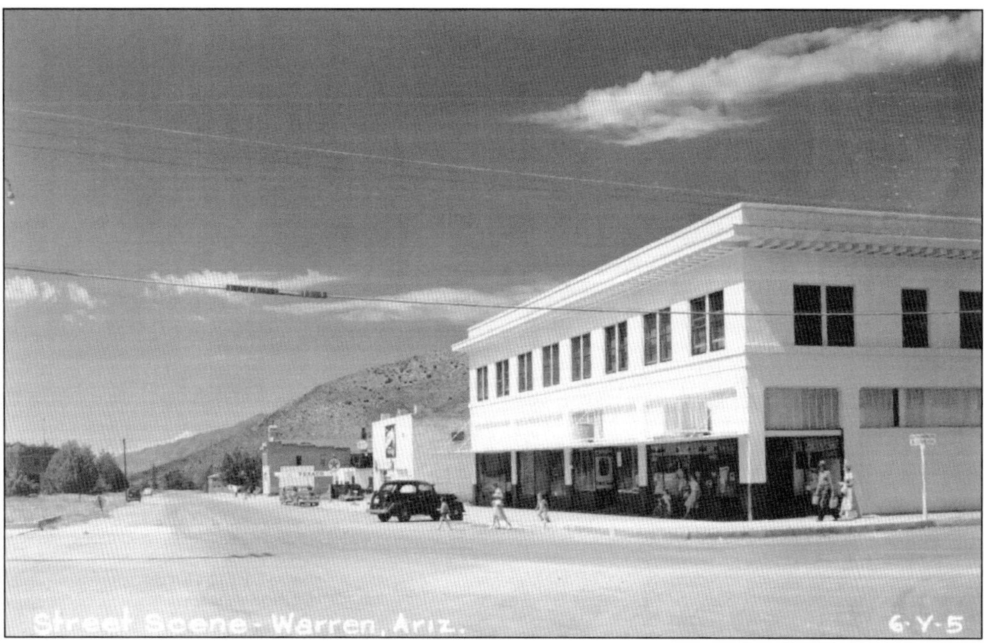

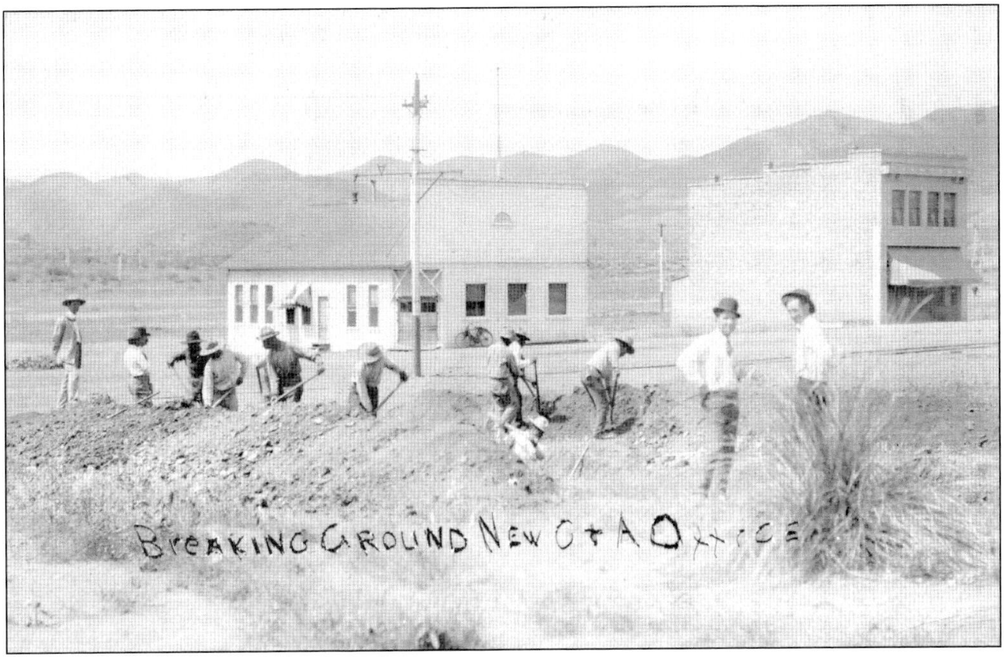

On the right of the above 1908 image is the first large structure erected in Warren, the office of the Warren Company. The postcard also captured the ground breaking of the Calumet & Arizona Mining Company's office building. Many years after the merger between the Calumet & Arizona and Phelps Dodge, the edifice went on to house the offices for the mining company's department heads. Shortly after the mines closed in 1977, the building became home to Bisbee City Hall.

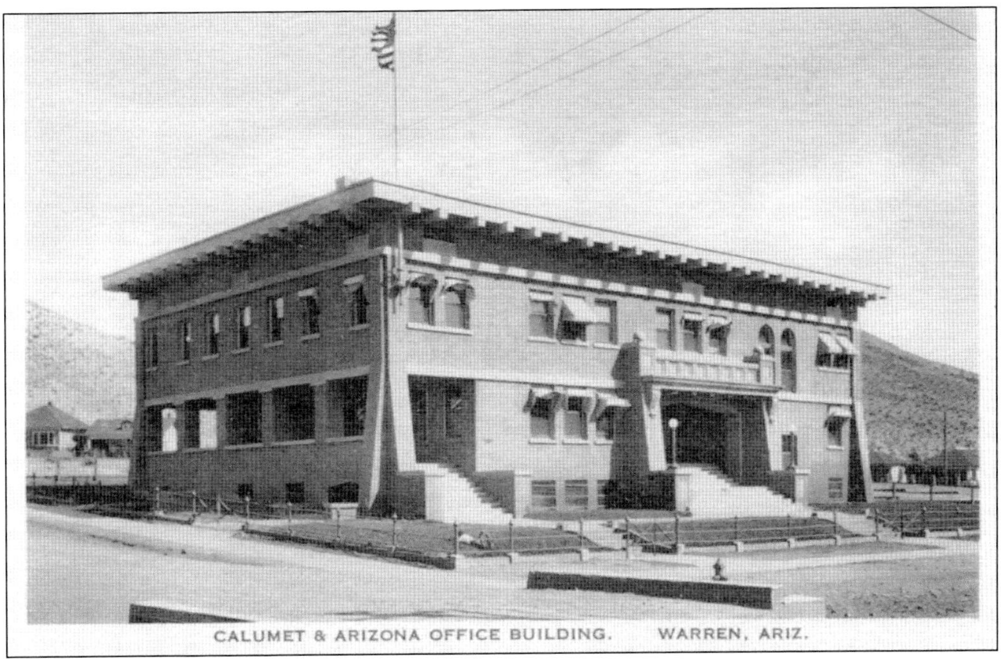

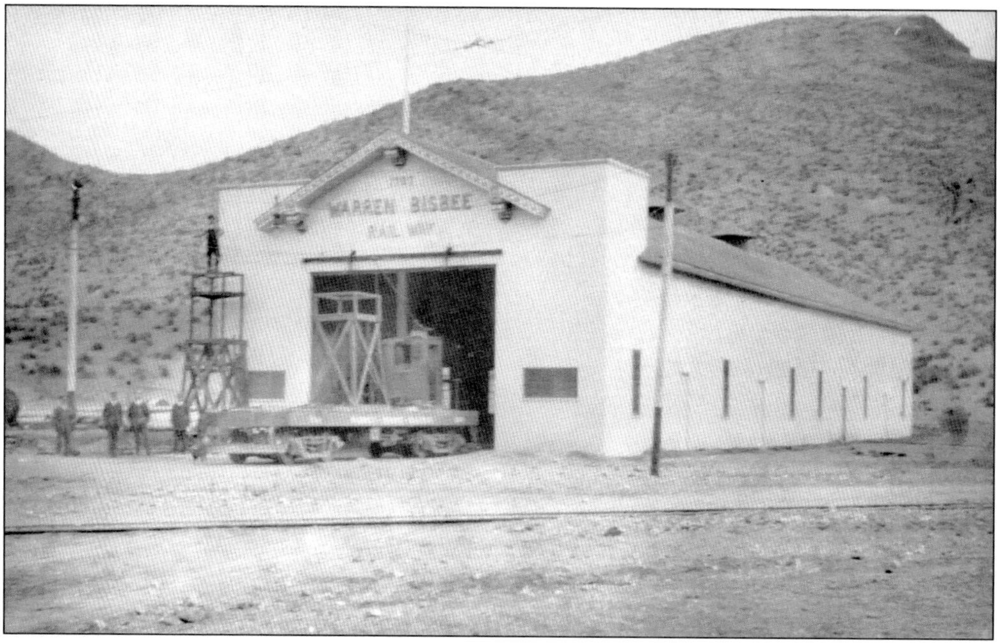

Due to the limited transportation options residents were presented with in the mining district, the Warren Company understood that a transit system had to be established. The primary commercial districts, as well as the majority of the mines, were located miles away, a walk neither miners nor their family members desired to make. Development of the Warren-Bisbee Railway began in 1907, and the Warren Company placed the system's maintenance car barn on Black Knob View, as seen above about 1908. Within several months the line connected Warren to Lowell, and by the end of the year, an estimated one million passengers had taken advantage of the district's new transportation system.

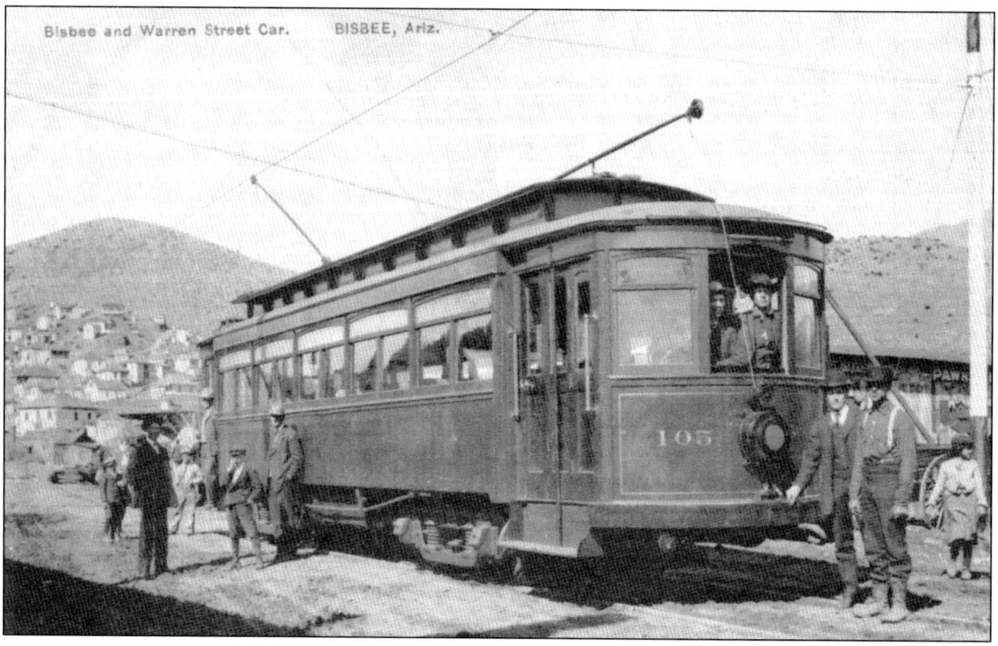

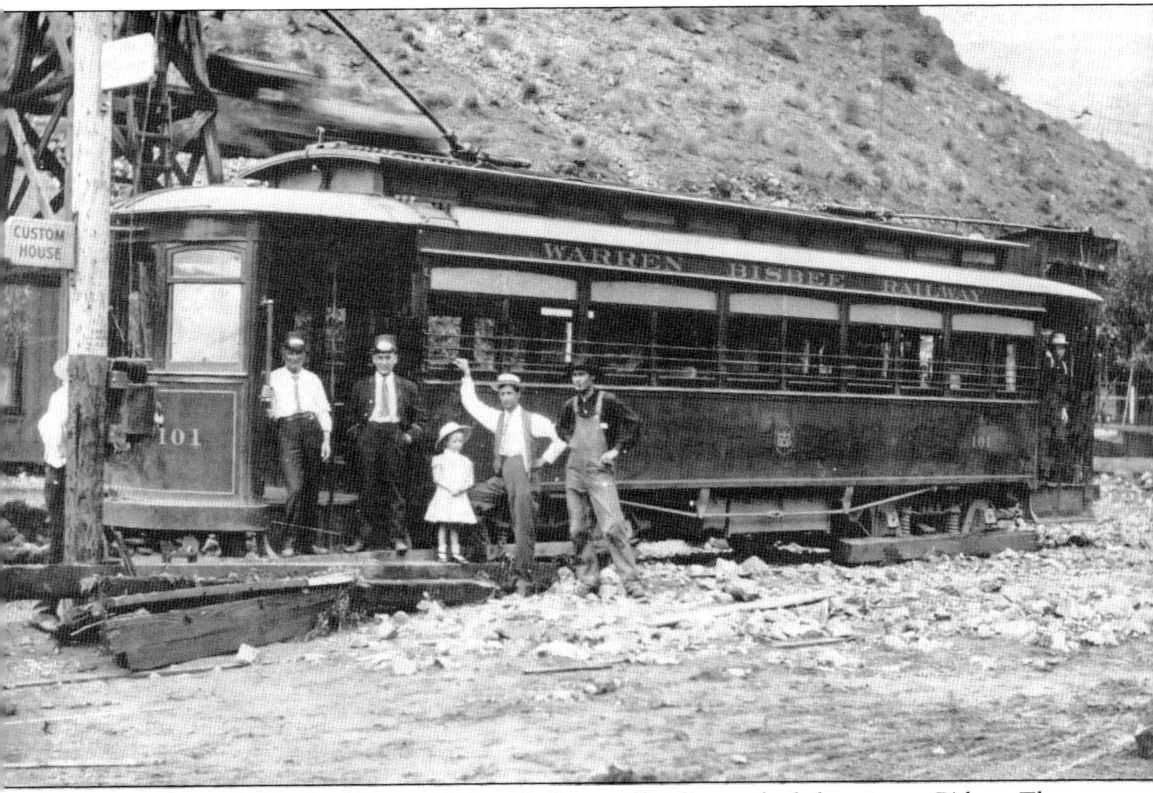

On March 11, 1908, the Warren-Bisbee Railway officially reached downtown Bisbee. The mining district's major communities were now all linked through the transit system. This was a momentous day in the area's history and was a cause for celebration. Thousands of residents flocked to Bisbee to witness the event first hand, with some even climbing telegraph poles to have the best view of the occasion. The mayor of Bisbee requested businesses to close during the event, and children were dismissed from school to take part in the historic day. The *Bisbee Daily Review* reported the celebration was "undoubtedly the greatest ever seen in the territory." As the streetcar approached the city, the Czar Mine's siren wailed, "dynamite salutes were fired in the old Copper Queen Mine," train whistles blew, the crowd cheered, and a mining company band played "America." It was official; the Warren Mining District was now a true metropolitan center in the American West.

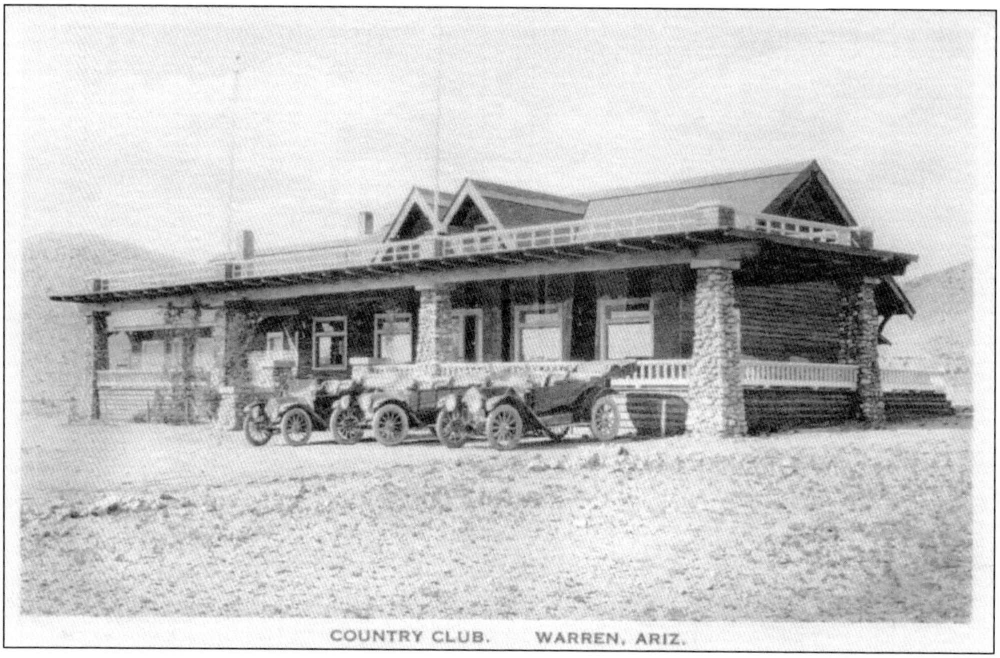

The southernmost point of the streetcar system was Warren District Country Club, which was established in 1908. Located about a mile south of Warren, the country club was the area's first and had 200 members within a year. The organization was yet another sign of the growing mining district's success and offered members opportunities for trap shooting, tennis, and also golf. The country club also hosted numerous social events, such as dances that were well attended.

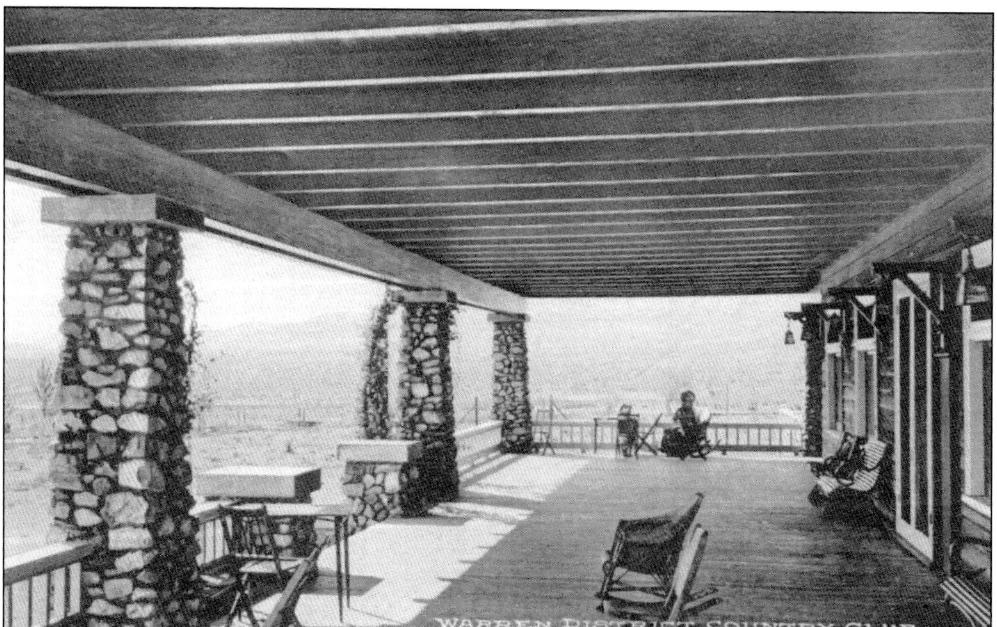

Four

Bisbee Life

Life was harsh in this desolate high desert. Those who lived in Bisbee lived in the moment. Justice was often mercilessly served at the hands of residents, and money flowed freely at the gaming tables, saloons, and brothels. However, as the face of the community evolved, it could see a future deeper than the bottom of a shot glass. Families were now becoming the foundation of the community, and the Wild West characteristics became more a part of Bisbee's historical past. Yet the mining camp continued to retain a distinct flavor, particularly during holiday celebrations. The Fourth of July was a glorious time in Bisbee, as the events drew families to the streets to join the festivities.

Another unique feature of Bisbee was the presence of the ever-faithful burro. These animals played a vital role in the mining camp, delivering water, goods, and supplies to the homes of residents perilously perched on the hillsides. Yet sometimes these burros carried an unwanted passenger. A lack of proper sanitary conditions in Bisbee proved to be a breeding ground for bacteria, which often made its way into the city's water supply.

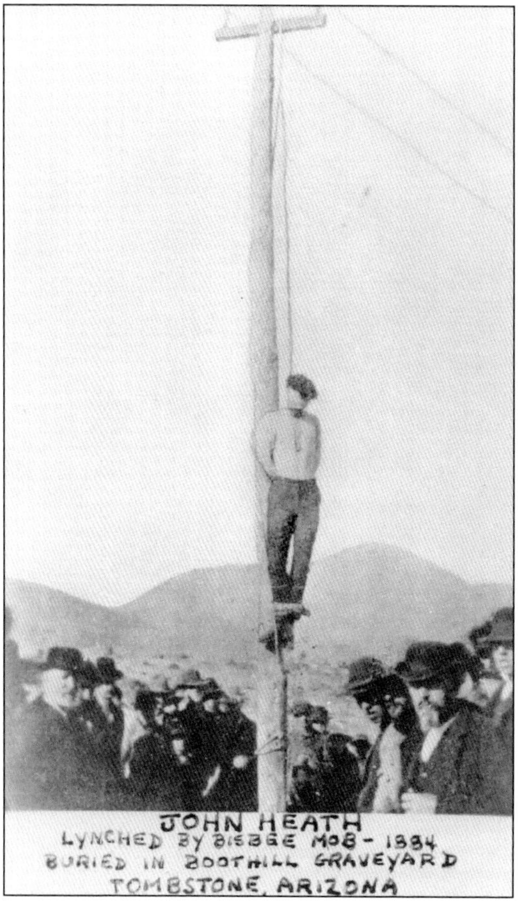

Despite being decades removed from the lawlessness of the late 19th century, postcards were produced featuring some of the violent incidents of the area's past. This image depicts the lynching of John Heith, who was implicated in the Bisbee Massacre of 1883. When Heith was not sentenced to death for his role in the incident, Bisbee vigilantes took him from the Tombstone jail and lynched him.

While many miners often chose a transient lifestyle, they could always count on one common feature in mining camps, a saloon. Saloons were plentiful in Bisbee, and one did not have to travel far up Main Street or Brewery Gulch to have a drink and engage in a game of chance. This image captured gamblers playing faro at the Orient Saloon in 1903.

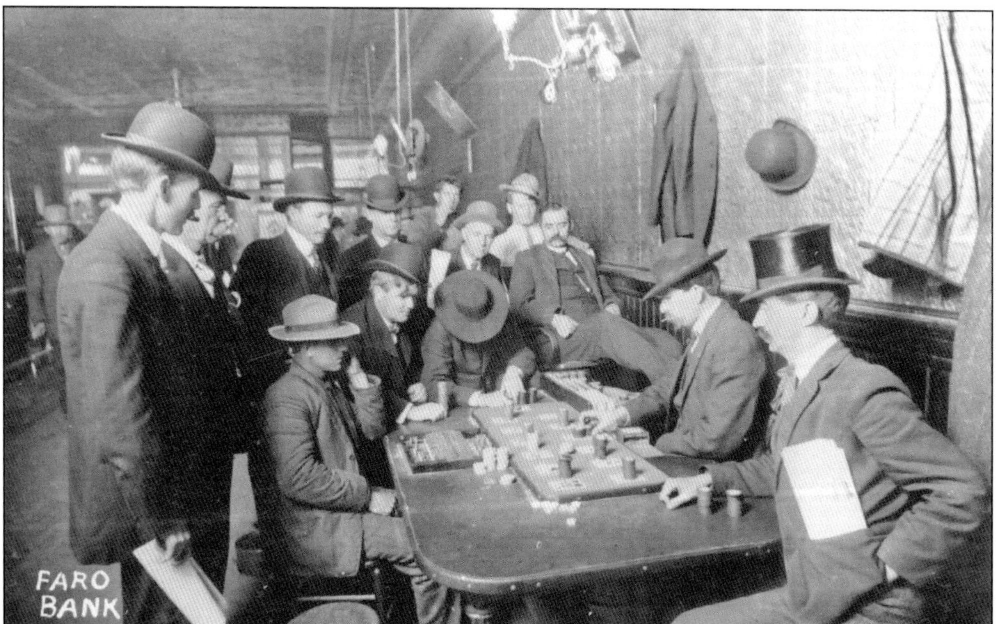

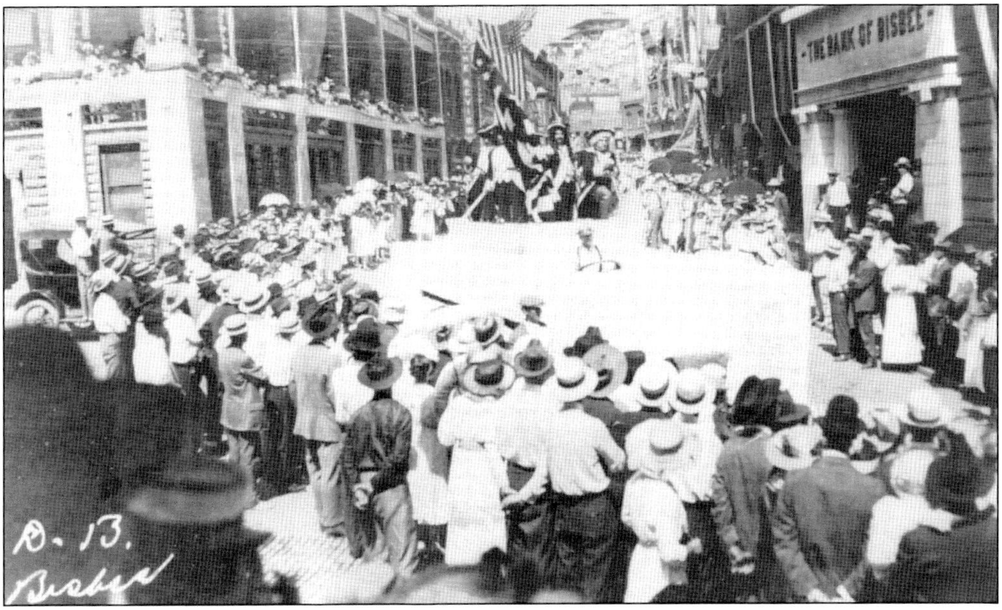

During the early 20th century, the Fourth of July was a time of great celebration in the Warren Mining District. Bisbee was often festively decorated with flags, banners, and bunting, and residents emerged onto the streets wearing their finest attire. One of the best attended events during the holiday celebrations was the parade. Thousands of people made their way to Main Street to witness the occasion, which often featured members of the military, decorative floats, and music from mining company bands that echoed through the canyons. The late 1910s image above captures a float surrounded by numerous spectators as it makes its way down Main Street. Below is the Fourth of July parade of 1909, which featured members of the 8th Cavalry who were stationed at nearby Fort Huachuca.

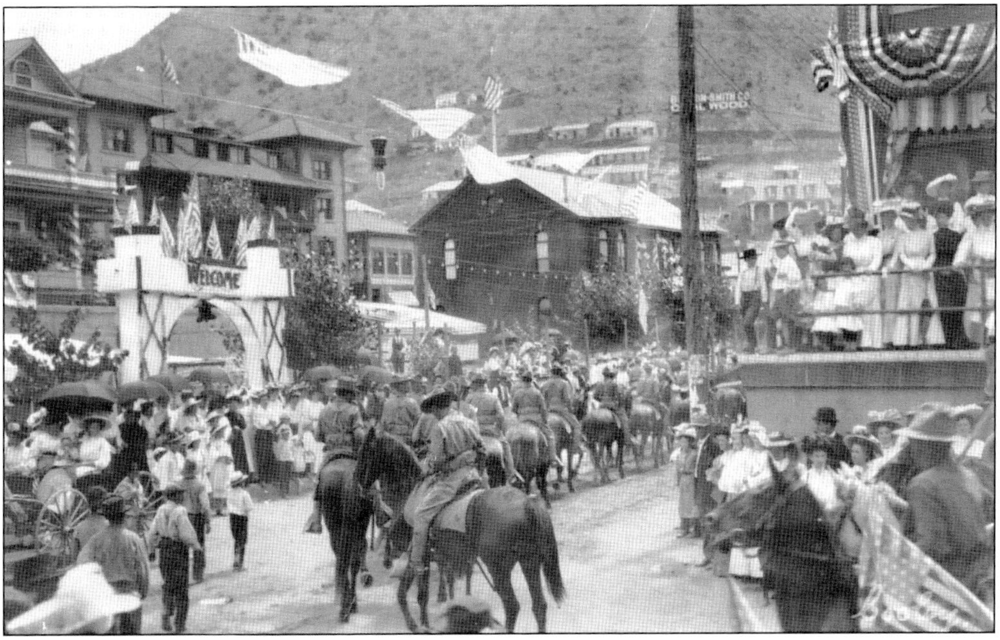

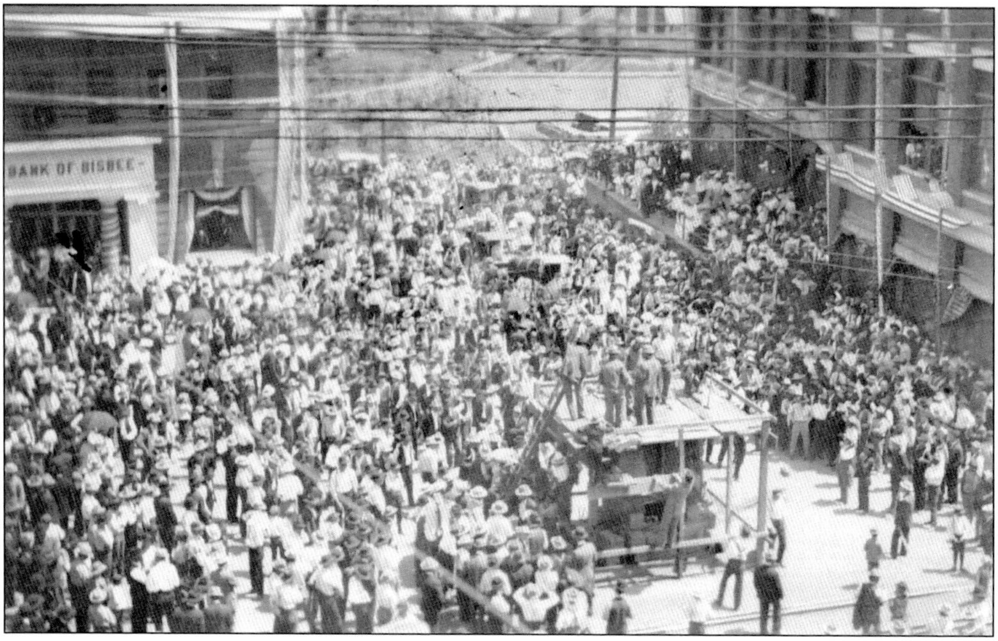

The Fourth of July events were a perfect blend of Bisbee's patriotism and mining heritage. Contests took place throughout the day, many of which were mining related, testing the skills of the district's miners. The most popular of these was the drilling contest, shown above in 1907 and below about 1910. During this competition, two men use a hammer and hand drillsteels to work away at a block of Gunnison granite, attempting to drill the deepest hole. These contests were quite popular in Western mining camps, and competitors traveled from all over the United States to take part in the event. Bisbee was proud of the records set at its competition, and residents were known to place hefty wagers on their favorite teams.

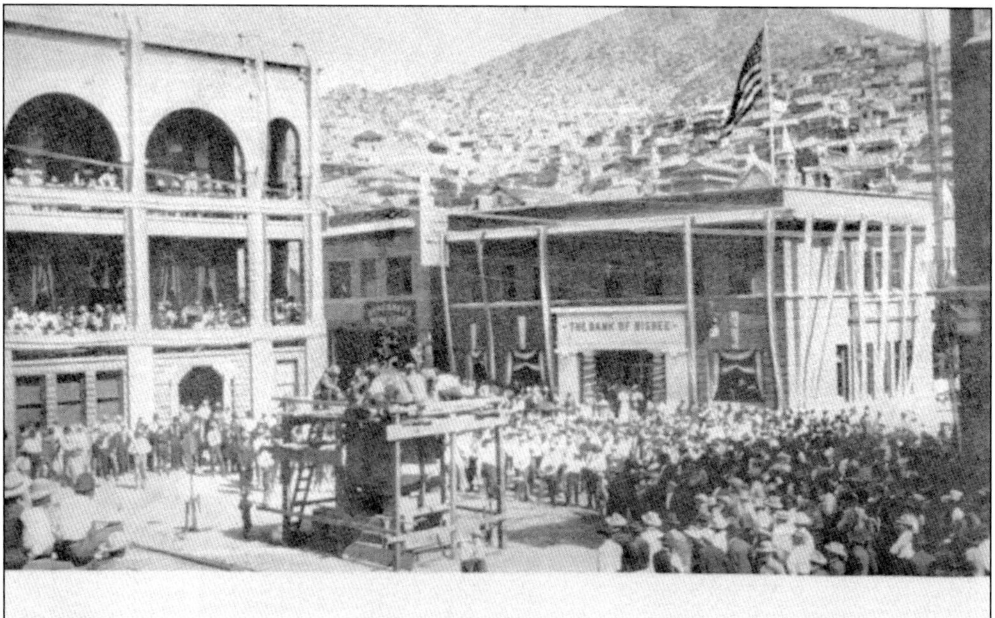

Drilling Contest, Bisbee, Ariz.

While mules provided invaluable labor within the mines, burros were the animal Bisbeeites looked to above ground. Able to carry much greater loads than people and traverse the steep and rocky mountains of the mining camp, these animals were used to transport a variety of goods and supplies throughout the community. Prospectors often looked to these trusted animals to haul their equipment while looking for and working at mining claims, as seen at right in a 1907 postcard. Additionally, burros were utilized in transporting ore from small areas of mineral deposition to the trains for shipment. The below 1910s image captured a burro pack train making its way through a mountain trail with valuable ore strapped to the backs of the animals.

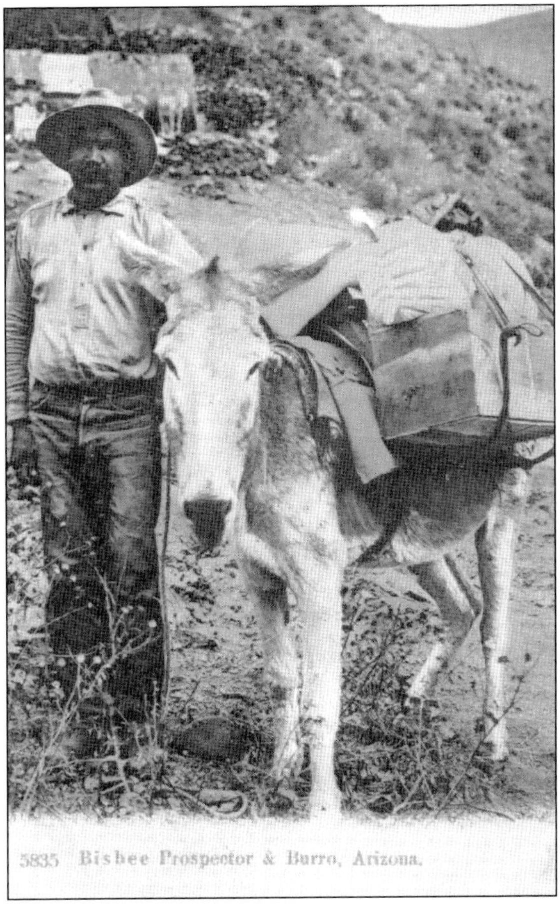
5835 Bisbee Prospector & Burro, Arizona.

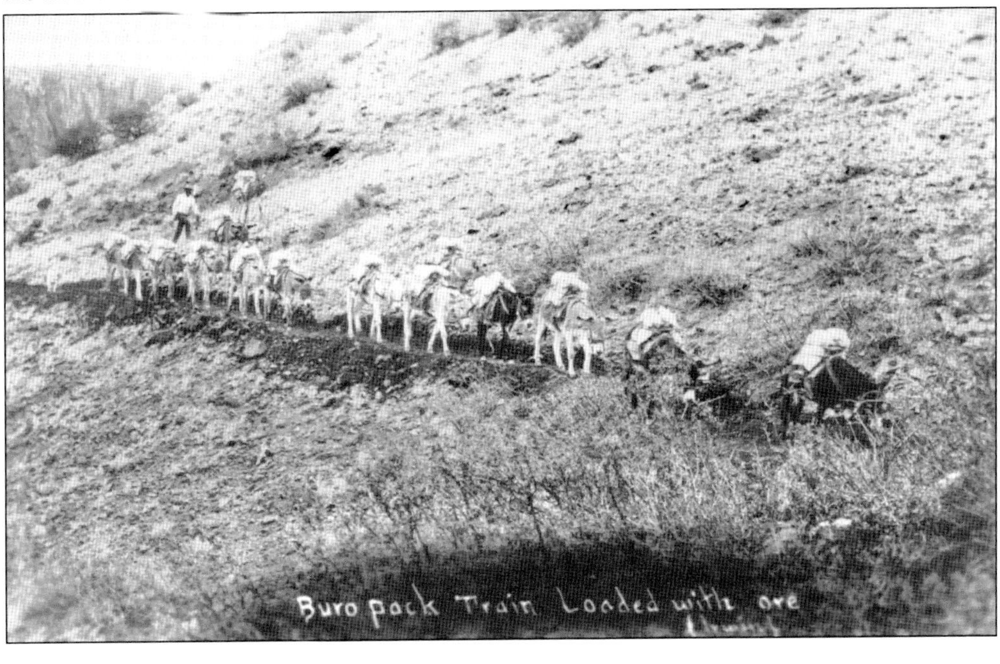
Buro pack Train Loaded with ore

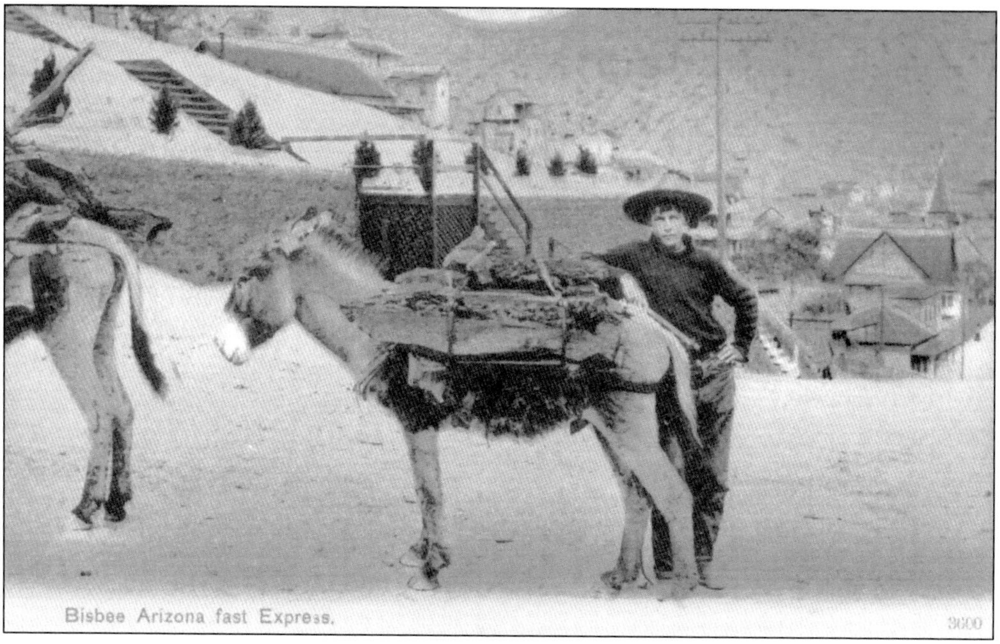

Bisbee Arizona fast Express.

3000

Bisbee is said to have been an area full of juniper and oak trees before people settled in the Mule Mountains. However, the necessity of wood for the mining process, coupled with material needs of the residents, resulted in the mountains soon becoming stripped of their trees. Woodcutters traveled within the canyons of Bisbee, removing trees and securing the wood onto the backs of burros. The wood would then be delivered to the homes of customers who used it for cooking and heating. These images from the 1900s and 1910s show the burros as they provided their invaluable service to the community.

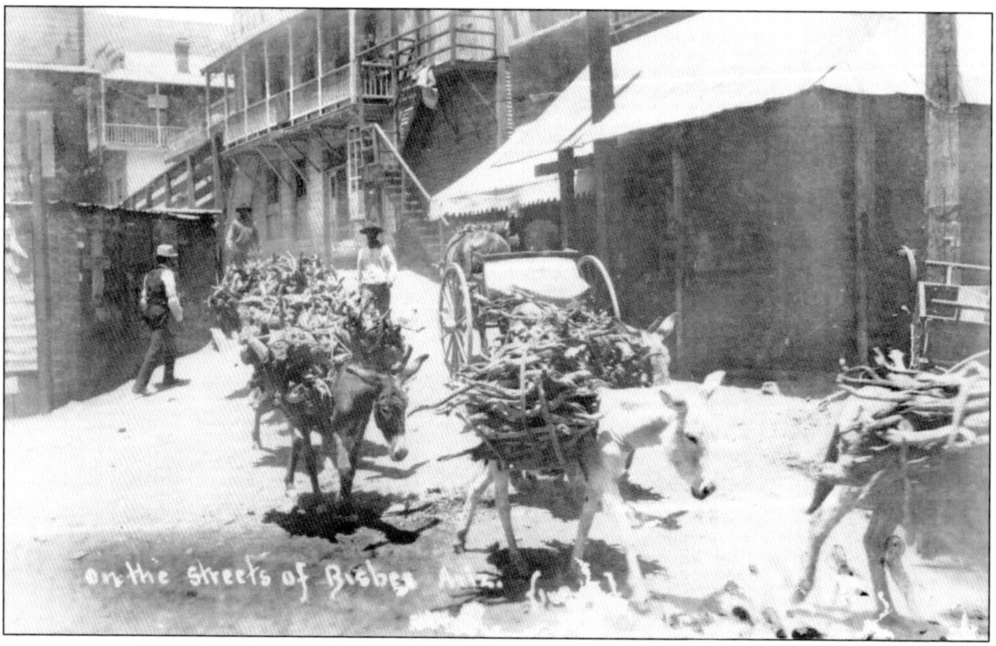

On the streets of Bisbee Ariz.

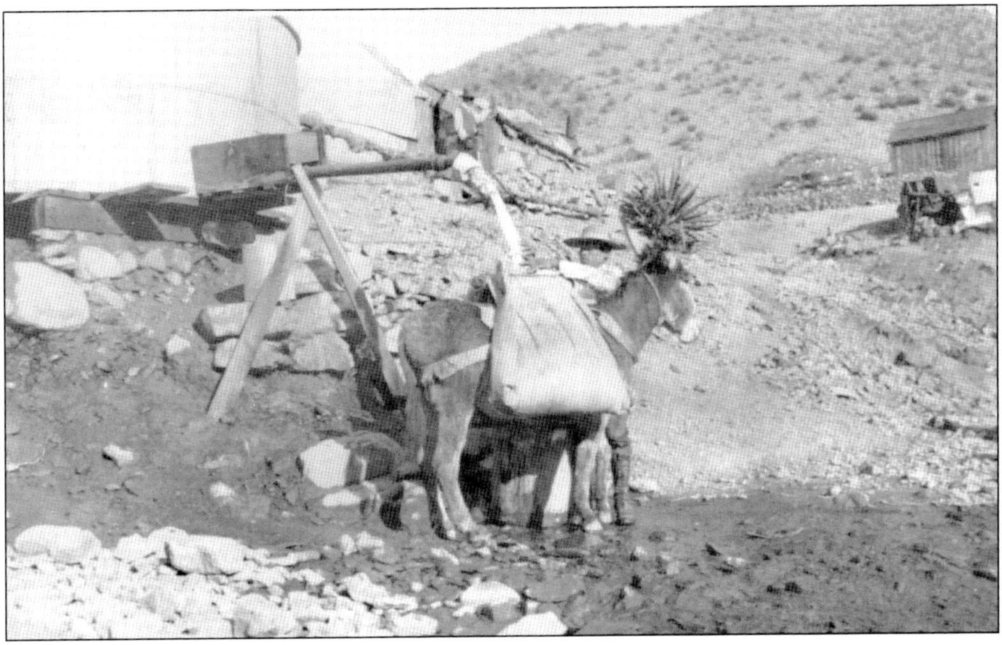

Burros were also used to carry the desert's most precious resource—water. Before residents in Bisbee were equipped with water piped into their homes, individuals purchased water from well owners. Water was filled into canvas bags that hung from the sides of burros, as seen above in 1908, and then delivered to the homes of residents. This was an expensive necessity in the mining camp and cost 50¢ per bag. When water was delivered, residents filled everything that could hold the valuable liquid, including barrels and large cans.

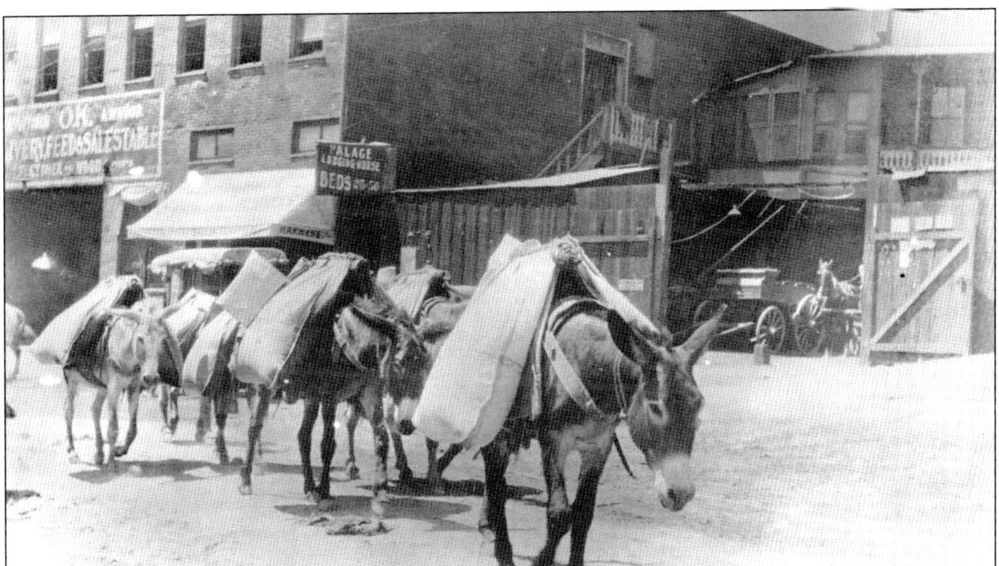

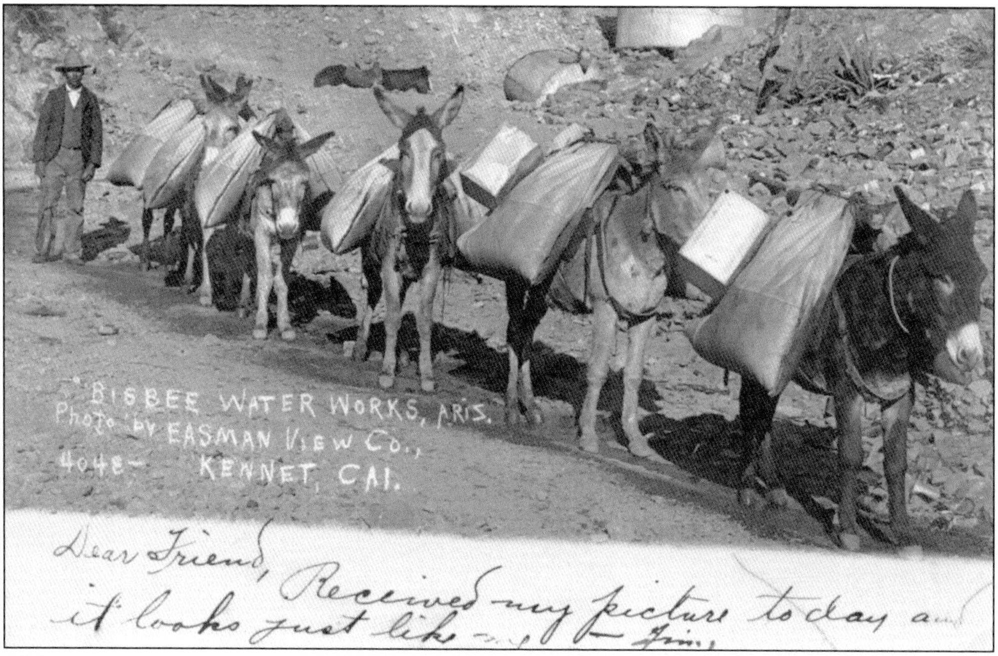

Although water was the source of life in the desert, it was also known to carry sickness and death in Bisbee. Before the mining camp witnessed community-wide improvements, such as sewer and water pipes and trash removal, Bisbee lacked proper sanitation. Consequently, serious illnesses such as typhoid fever became a major problem within the mining camp, with one of the factors being contaminated water. Some of Bisbee's regularly used water supplies were located downhill from polluted ground, including the cemetery and outhouses. The postcard below shows a man filling the canvas bags on a burro from a water tank. However, to the left, directly uphill from the tank, is an outhouse that held the potential to contaminate the water with bacteria.

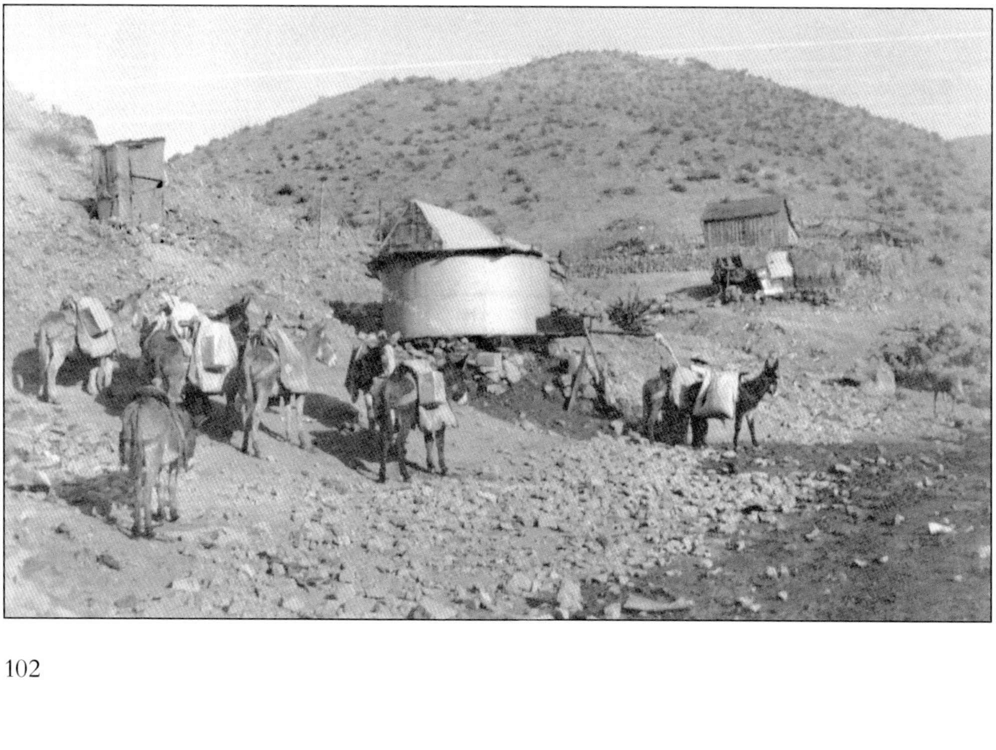

Five

Floods and Flames

Just as the ore deposits within the Mule Mountains influenced how the Warren Mining District developed, so did the disasters that fell upon the region. Like so many other communities in the early 20th century, Bisbee was wrought by deadly floods and devastating fires. Many of these were captured by local photographers who produced postcards of the alarming events, which acted as a means of visually communicating the region's news to the cards' distant recipients. The images shed light on the horrific experiences that residents endured in the mining camp, as well as their resilient nature. Bisbeeites were quick to rebuild Main Street from its fire-stricken skeletal remains, to clean up streets after a flood, and, most importantly, to provide assistance to disaster victims.

Residents of the mining camp learned early on to take precautions against these calamities. Through actions such as paving streets and creating an efficient channeling system for floodwaters, the community witnessed a decrease in the damage created by these torrents. Additionally, in an effort to prevent the rapid spread of fires in the city, Bisbee placed restrictions on the construction of wooden buildings within the city limits. These efforts not only increased safety within the city but also strengthened its foundation as a mining camp that was not going to bust anytime in the near future.

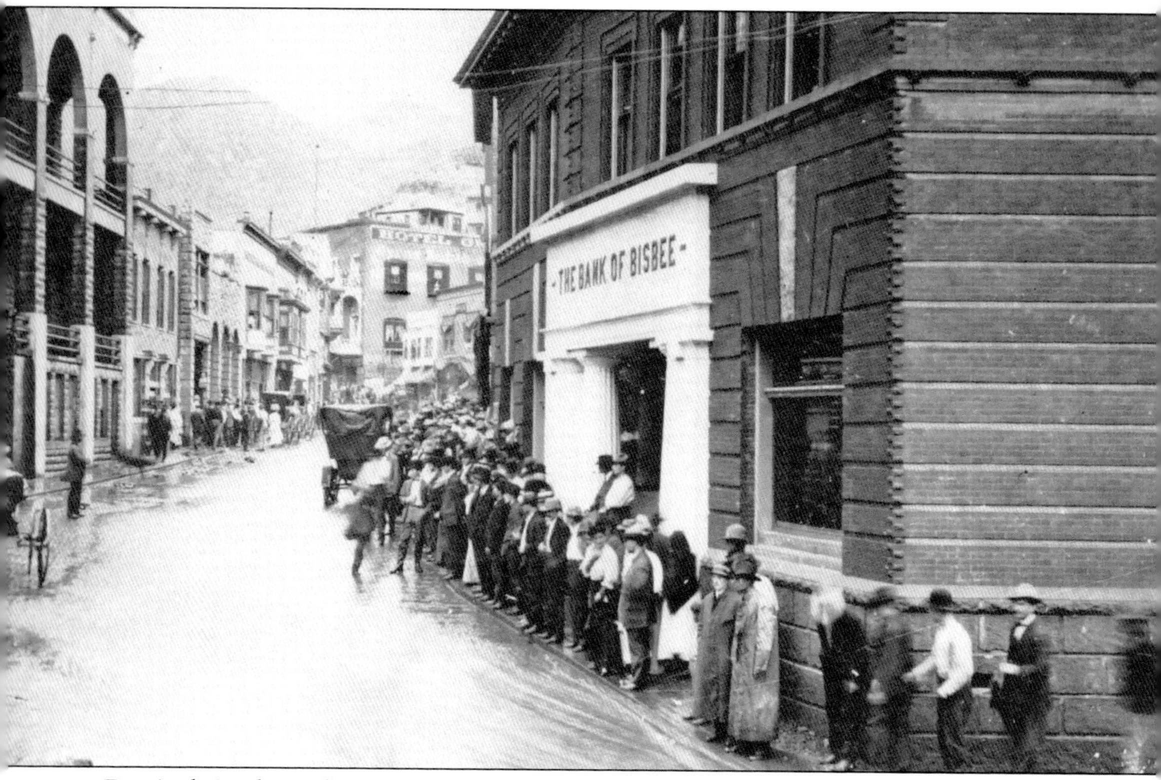

Despite being located in Arizona, a state known for its high temperatures, Bisbee's mile-high elevation within the Mule Mountains offers relief from the scorching summer heat. When the warmth reaches a point of undesirability, Mother Nature brings relief in the form of monsoon rain that quickly lowers the temperature. However, the sources that enable Bisbee's forgiving climate have also produced devastating results. The region's topography combined with the torrential summer rains provided the perfect recipe for flash floods to rage down the canyons of the mining camp. On numerous occasions these floods have swept through Bisbee taking no prisoners, destroying property, and killing animals and residents. Bisbeeites learned to both respect and fear the power of the floodwaters, never wishing to intrude upon their path.

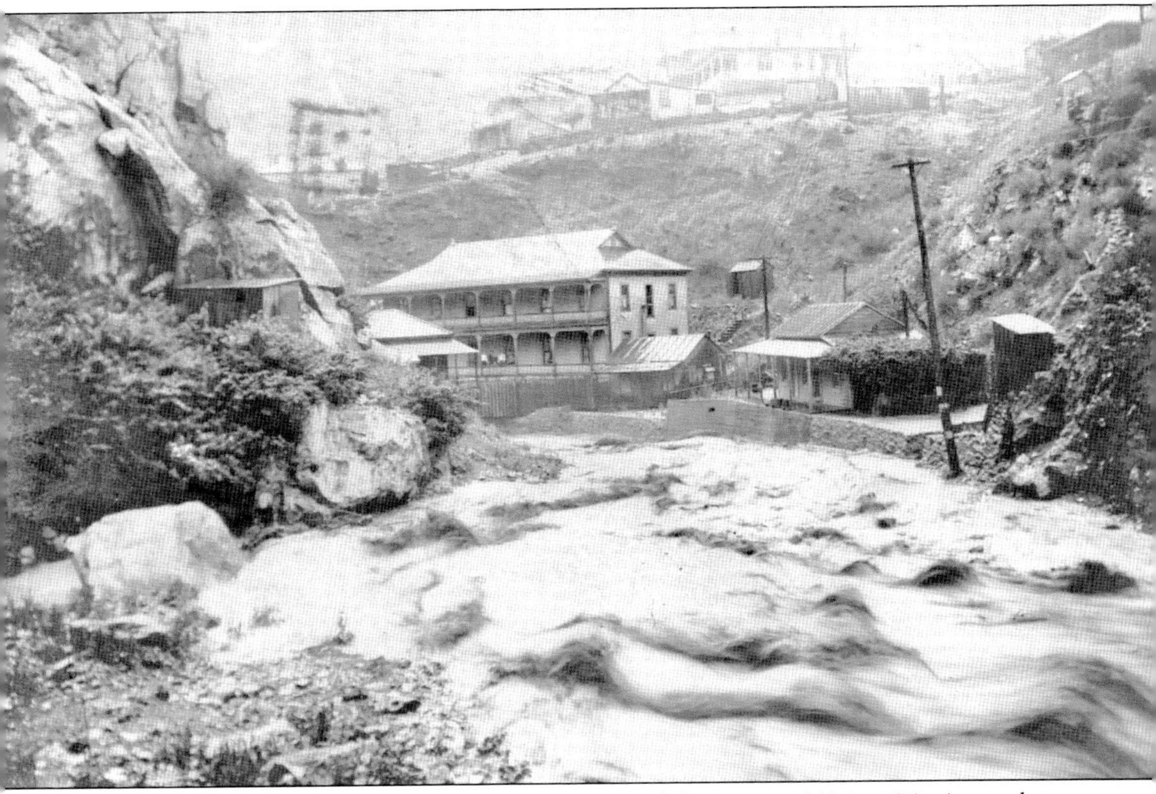

Over the years numerous floods overtook the streets of the Warren Mining District, such as this 1907 torrent at the base of Castle Rock. A sudden eruption of a cloudburst could result in a flash flood forming in minutes, leaving little time to retrieve vulnerable property or reach safe ground. The forceful nature of these waters was legendary to residents, who understood that if they dared to challenge the strength of a flood, it would likely win. In an effort to reduce the damaging effects of the waters, a ditch system, called the subway, was created. This system, coupled with the installation of floodgates, enabled water to be diverted behind busy Main Street away from businesses. These preventative efforts proved to be effective in channeling water away from Main Street.

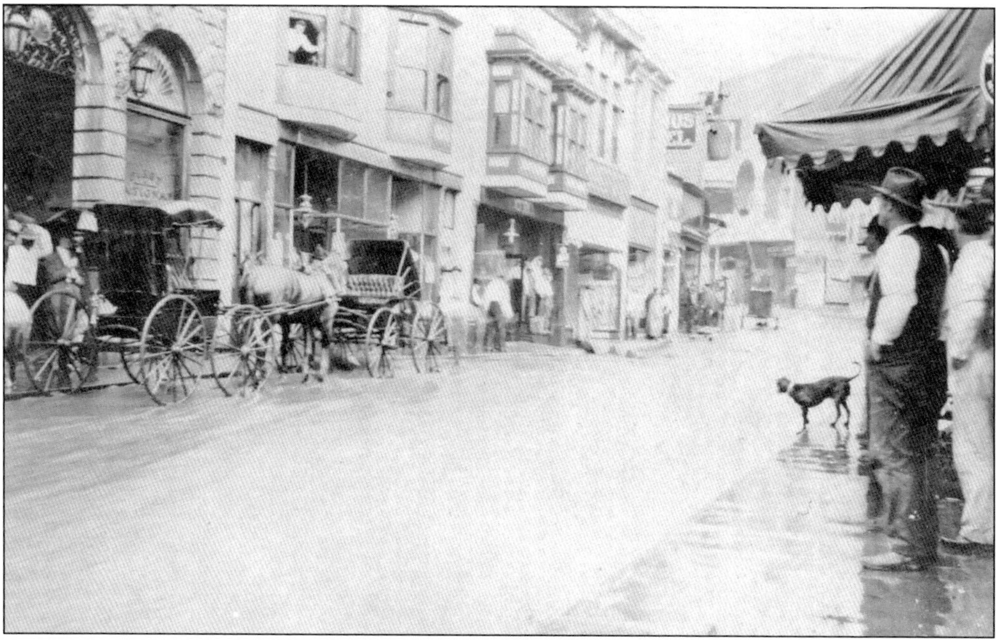

When floodwaters appeared, the lives of residents and business owners were at a standstill. With every storm came the possibility that they would become a victim of the violent rain. Streets quickly turned into waterways trapping unfortunate citizens who ventured out of their homes on the street, or some inside businesses. People had to fear not only the water but also the debris it carried and the potential landslides created by the deluge of rain. It was not rare to witness carriages or automobiles being carried off by the water or people being struck by rocks falling off a nearby mountain during a storm.

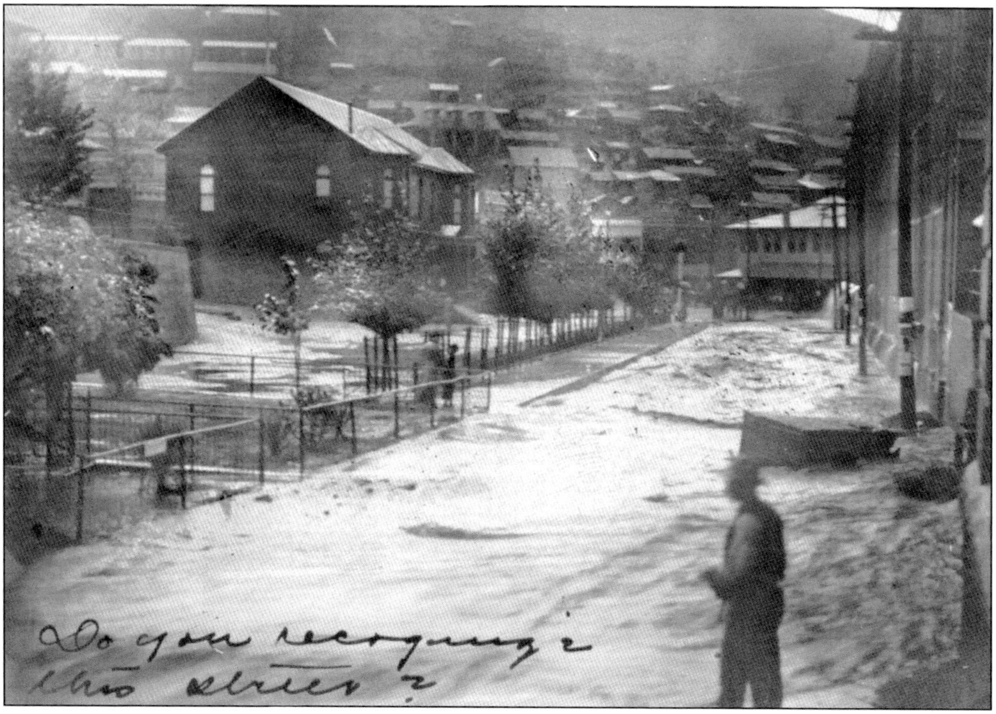

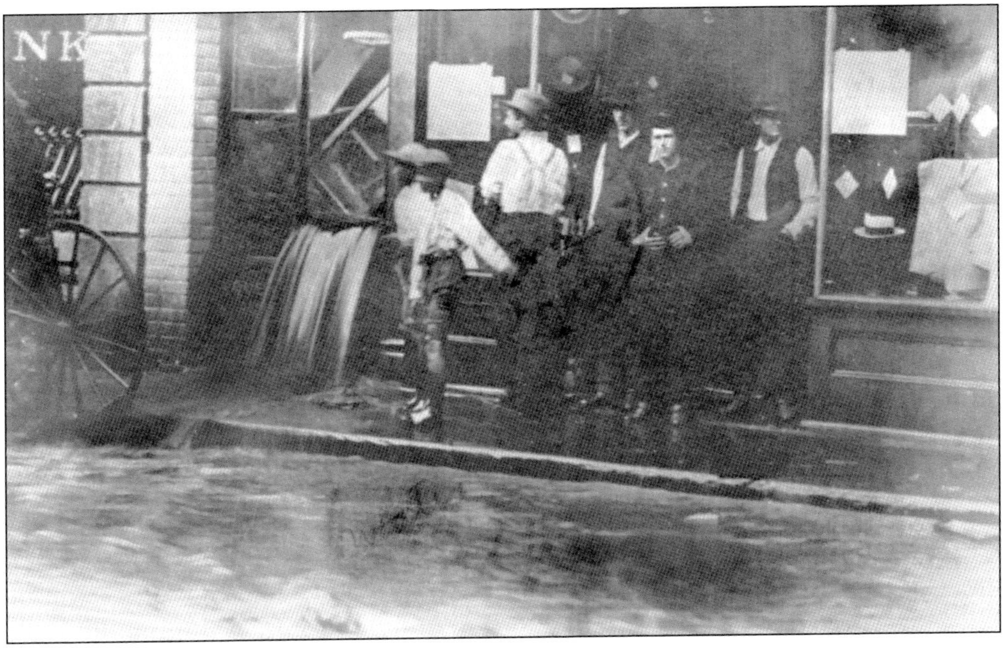

August 1908 proved to be a challenging year for Bisbee residents, as three violent cloudbursts erupted in the mining camp in three weeks. The torrents raged and were unforgiving to residents as well as business owners. Floodwaters flowed into establishments along Main Street, ruining and sweeping away merchandise, while leaving inches of rock and mud behind. The postcard above captured the flood of August 4, 1908, with water pouring out of a shop's broken window on the left. Below, the image shows water exiting out of the door of a Main Street building. Looking closer at the window that is second from the right, there are two people who are apparently standing on top of an object to escape the waters inside.

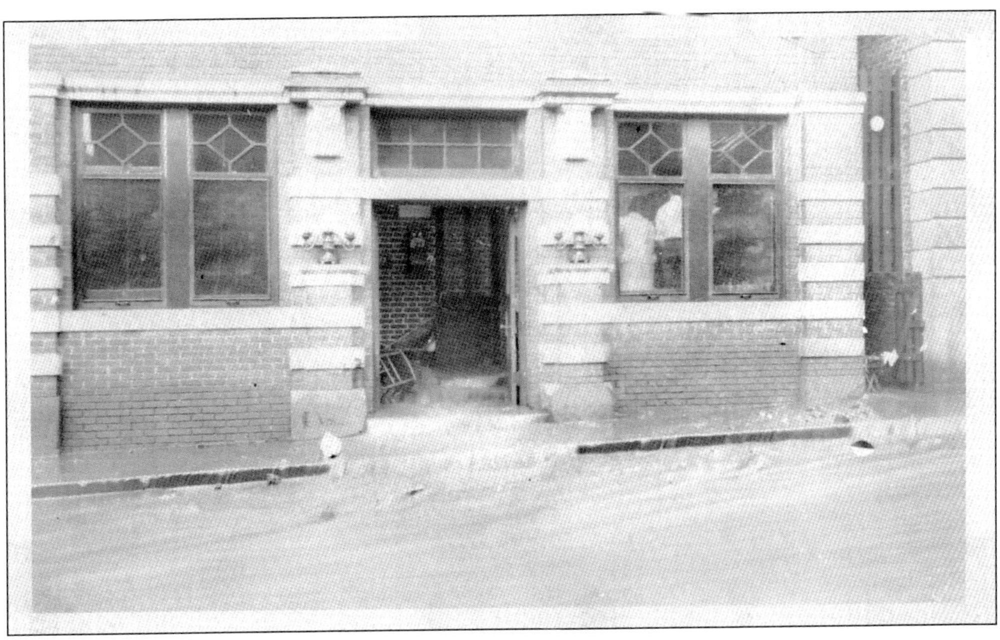

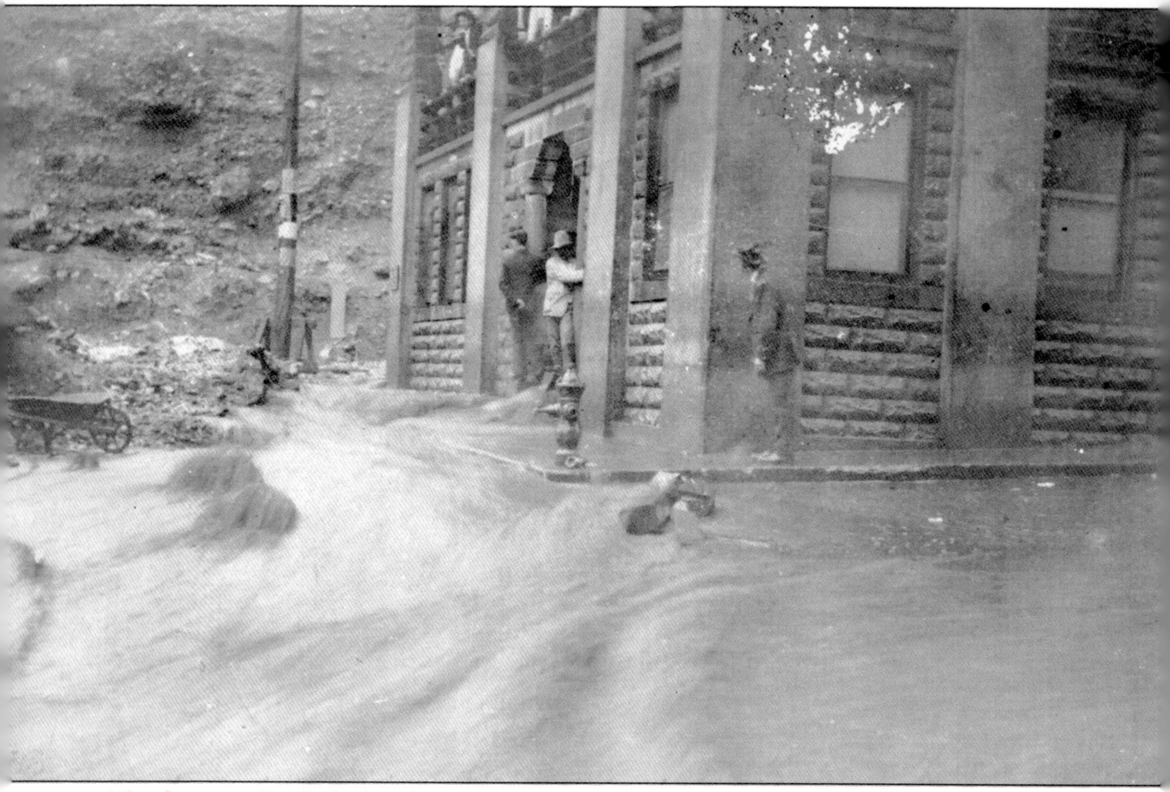

The first two floods during August were particularly devastating for the city's post office and businesses located on the south side of Main Street. The *Bisbee Daily Review* reported the storm of August 4, 1908, wrecked the post office, swamped the stores and saloons, and created gullies within roads that were up to five feet deep. Property damage caused by the storm was estimated between $15,000 and $20,000. The post office was noted to have been hit the heaviest by the flood, with the building's interior sustaining significant damage in addition to countless residents' mail being destroyed or swept away. Merely six days later on August 10, just as the post office was recovering from the last storm, another wreaked havoc in the area, flooding the building once again. This August 1908 image shows two men trying to escape the floodwaters by clinging to the pillars at the post office entryway.

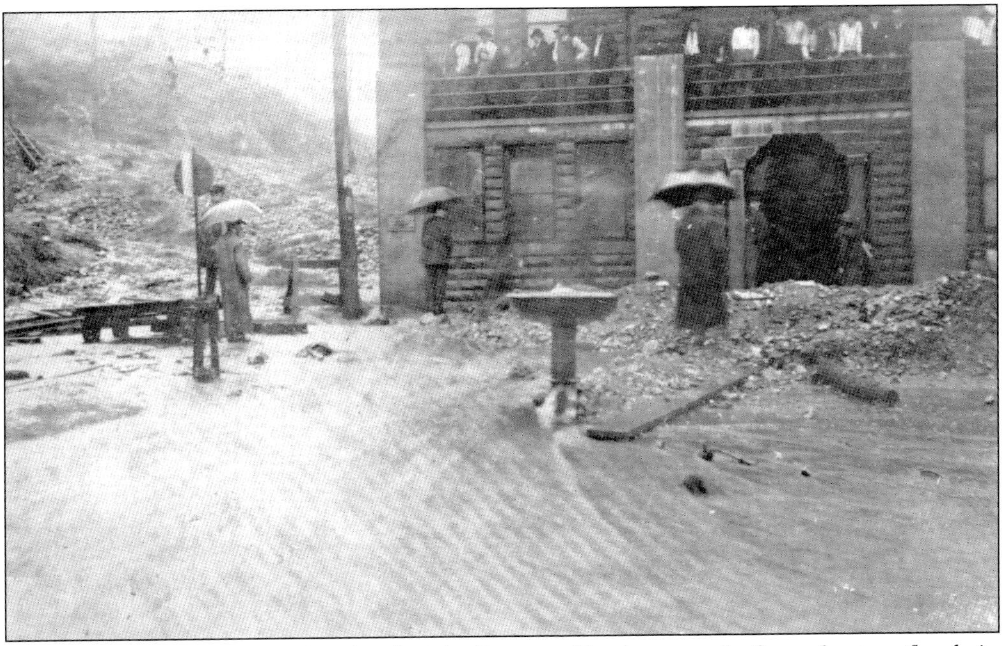

One of the reasons the south side of Main Street suffered so greatly from the two floods in early August was due to landslides that occurred behind the buildings on each occasion. These created a dam that channeled the water into the buildings. The *Bisbee Daily Review* stated during the flood of August 4, 1908, between 1,500 to 2,000 tons of rock were "piled by the water in less than ten minutes" behind Main Street. The post office suffered the greatest damage, and employees not only had to contend with the water but also boulders crashing through the windows of the building, leaving employees fleeing for safety. These images from early August 1908 reveal the immense amount of rock and debris left by the landslides behind Main Street.

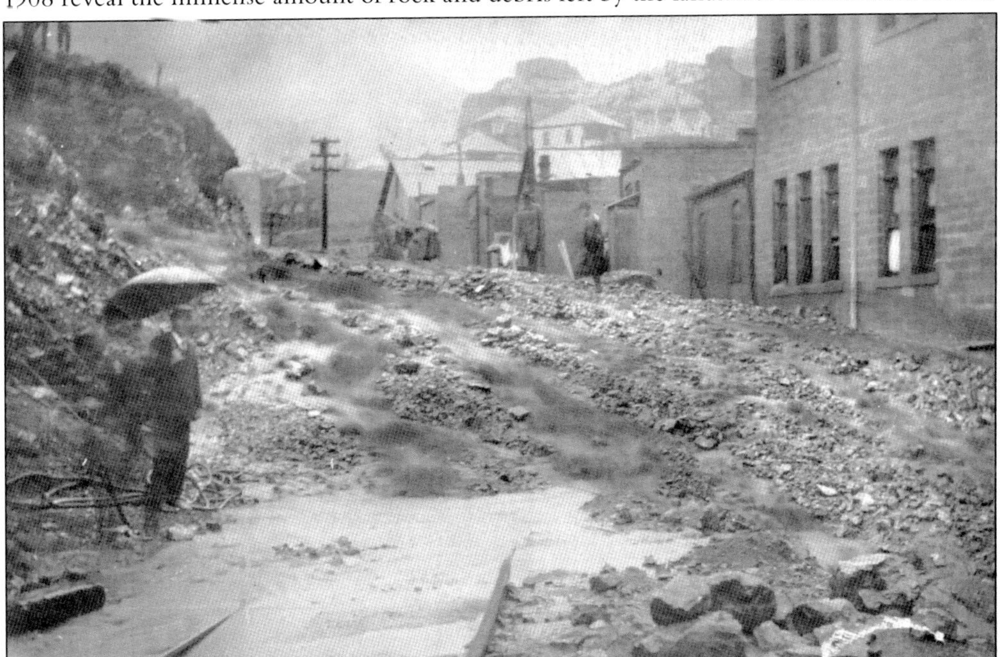

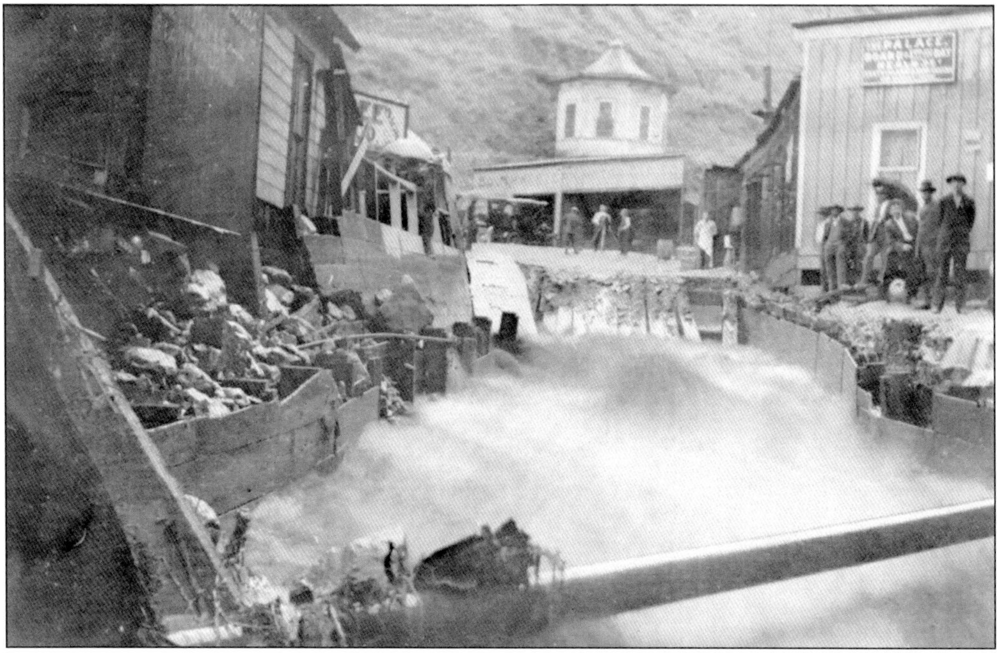

On August 11, 1908, the *Bisbee Daily Review* warned its readers that a violent flood was "liable to be repeated at any moment, with even more disastrous results." Unfortunately for the community, this would prove to be true before the month was over. On August 24, 1908, the subway was tested by one of the strongest floods in Bisbee's history and failed. The *Bisbee Daily Review* reported, "With the roar and velocity only attainable by a mountain torrent, the subway . . . gave way." Terror filled the hearts of individuals, and "as the long section of street above the underground waterway began to heave and quake, men and women fled for their lives, with pistol shots and cries they warned those below of the impending danger." Nothing could stop the wave of water that began racing down Main Street.

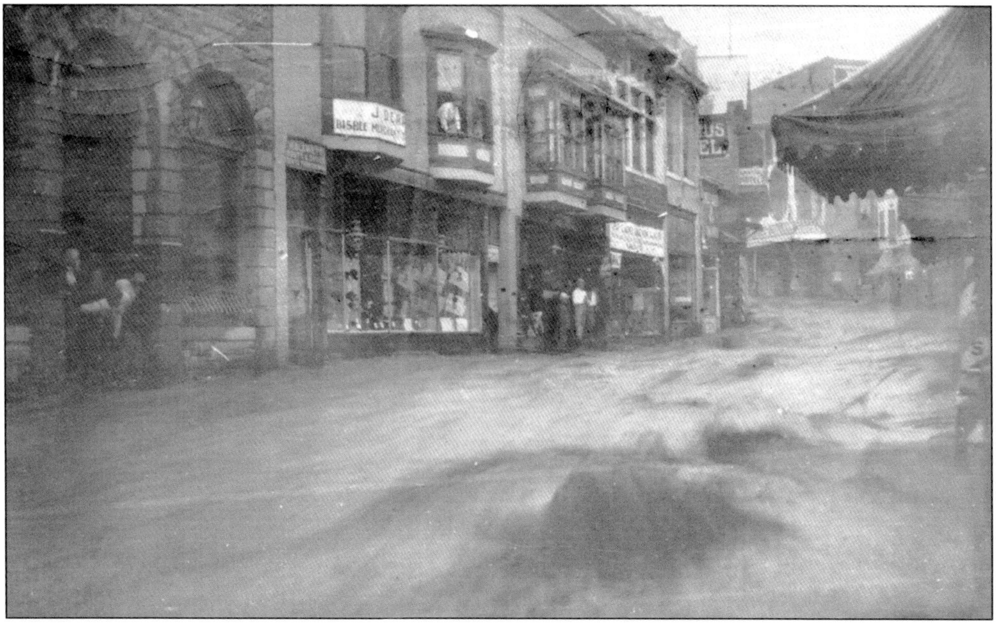

Within 15 minutes, "every street in the camp was impassable," leaving individuals stranded and wondering how other parts of the mining camp fared. An estimated four feet of water eventually overtook Main Street and made everything in the path of the undiscriminating torrent its victim. This included horses that were tied on posts and attached to buggies. Several individuals were successful in rescuing horses, but many residents could only watch helplessly as these steadfast animals were swept away by the deluge. When the floodwaters subsided Main Street was in ruins, with rock and debris from the failed subway piled in the road.

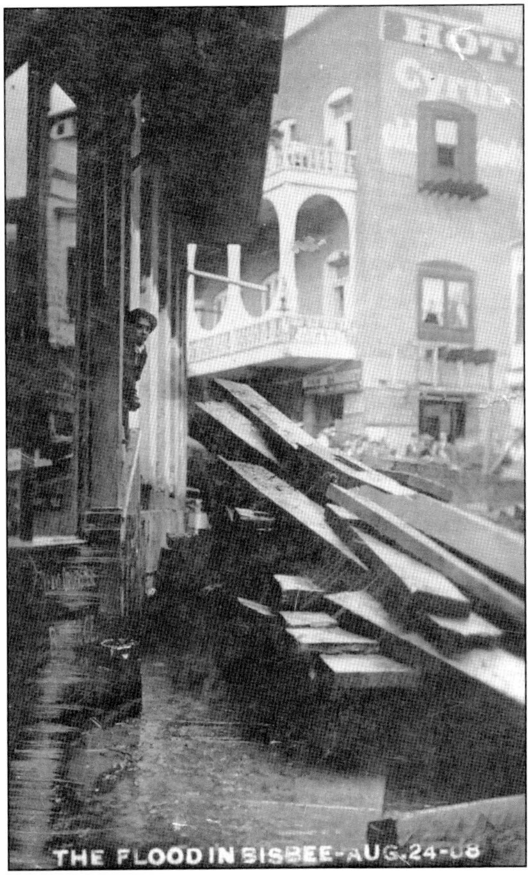

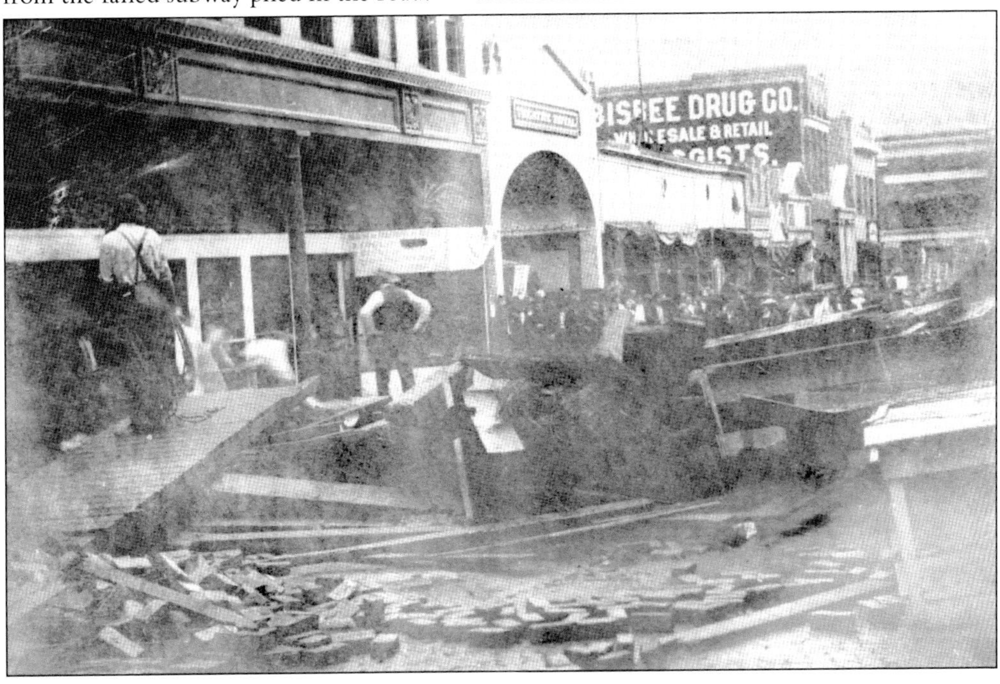

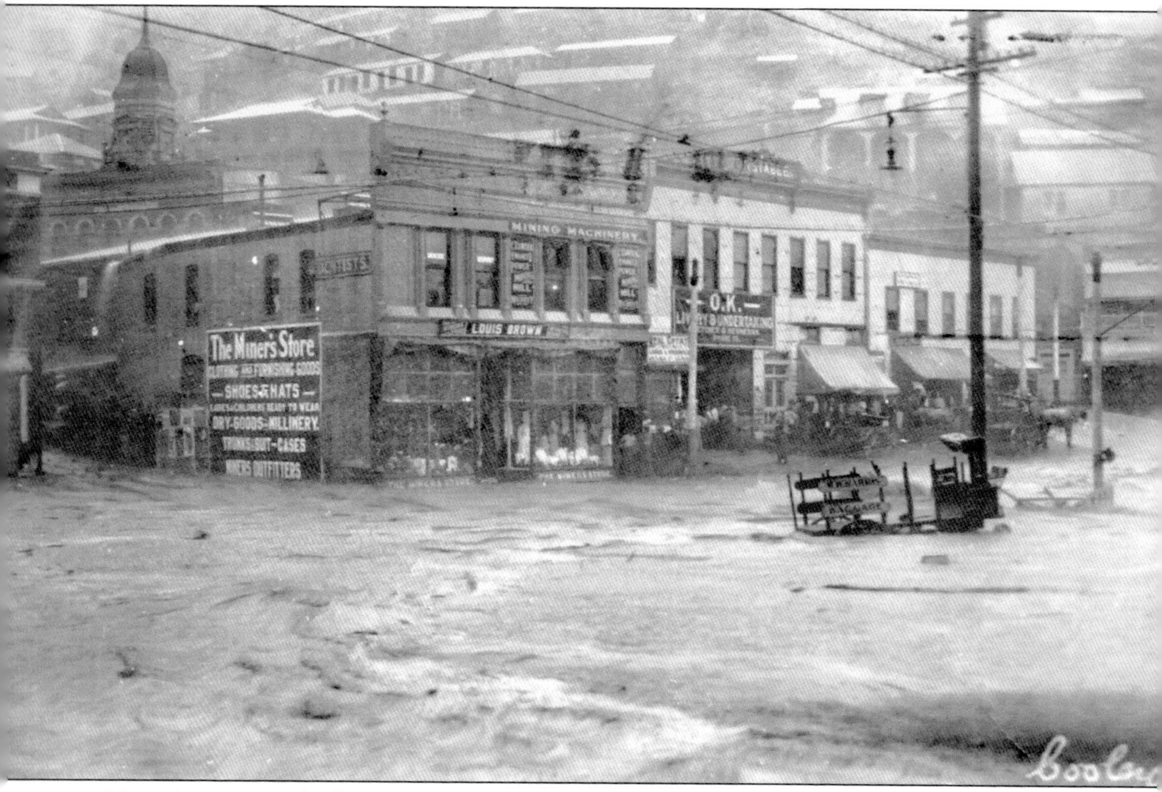

Never immune to the forces of Mother Nature, the community experienced another devastating flood on July 22, 1910. Three hours of continuous rain pelted the community, rapidly flowing down the mountainsides and creating deadly floodwaters within the canyons of the mining district. This image captures the flood at the intersection of Tombstone Canyon and Brewery Gulch where the waters violently merged. With little effort, one can see the direction the forceful water is taking, as well as the debris being carried down the canyons. The *Bisbee Daily Review* stated that the noise from cans, boxes, and beer kegs that flowed down the concrete lining of Brewery Gulch "added a new voice to the thunder of the stream." However, when the waters began to carry beds, furniture, and distinct portions of houses, residents began to closely watch the deluge for people who may have been swept away by the flood.

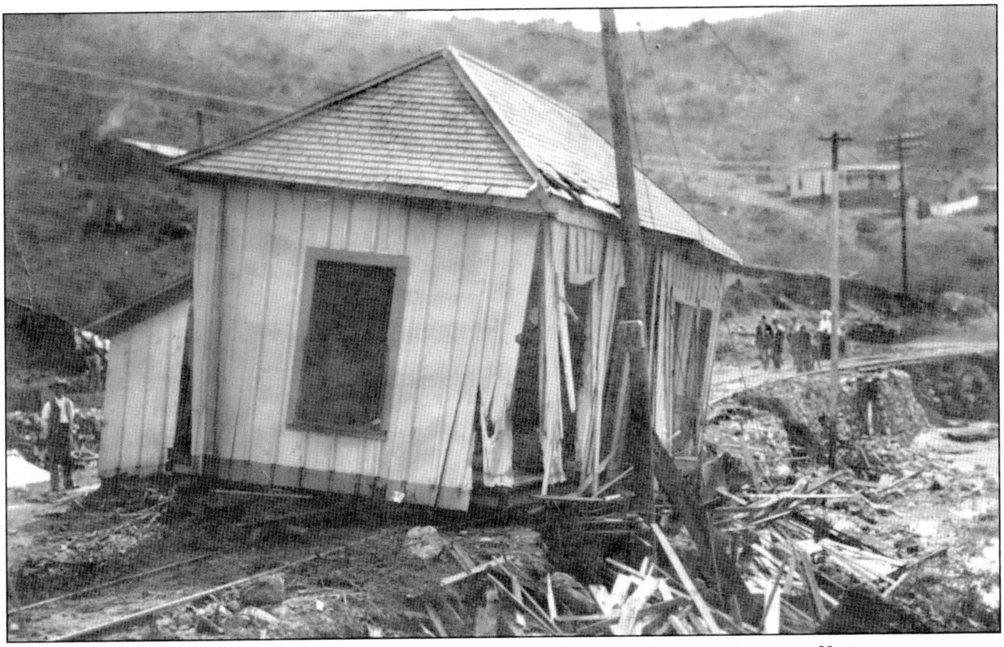

These powerful floodwaters left Bisbee in ruins with countless residents suffering some type of damage to their property. Much of the destruction took place between Bisbee and Lowell where roads were made into deep gullies and homes slid down hills. These images reveal the damage sustained by many homes in the district on that fateful day. Numerous families lost everything they owned and looked toward the kindness of neighbors to provide them with temporary shelter and supplies. The July 23, 1910, *Bisbee Daily Review* is filled with remarkable accounts of residents escaping certain death, as well as those who were not so lucky, having their lives extinguished by the merciless torrent. Two people died as a result of the flood, both trapped inside buildings that the flood swept away.

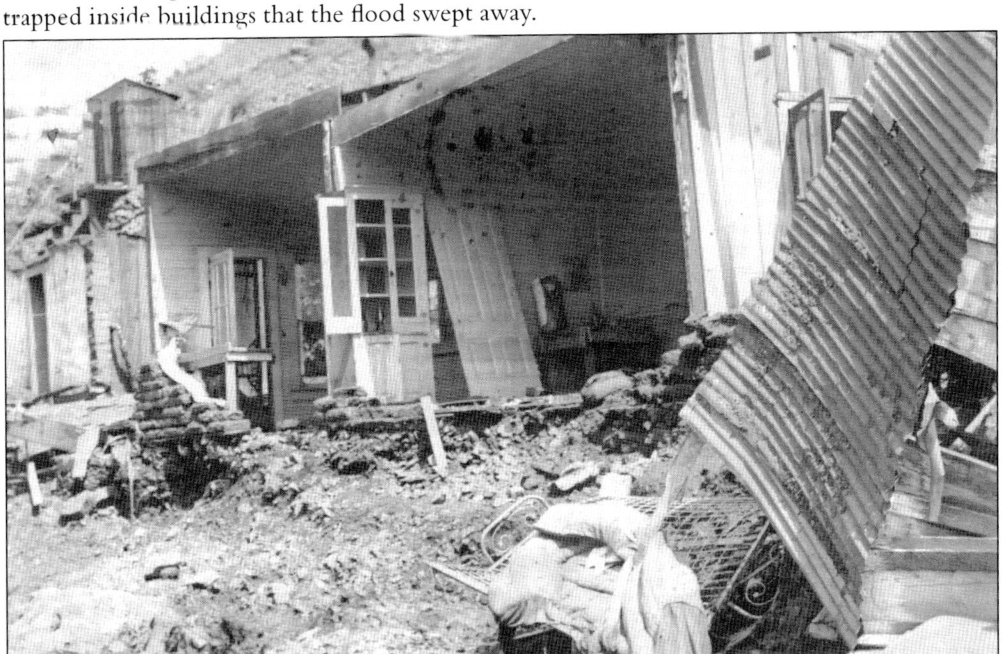

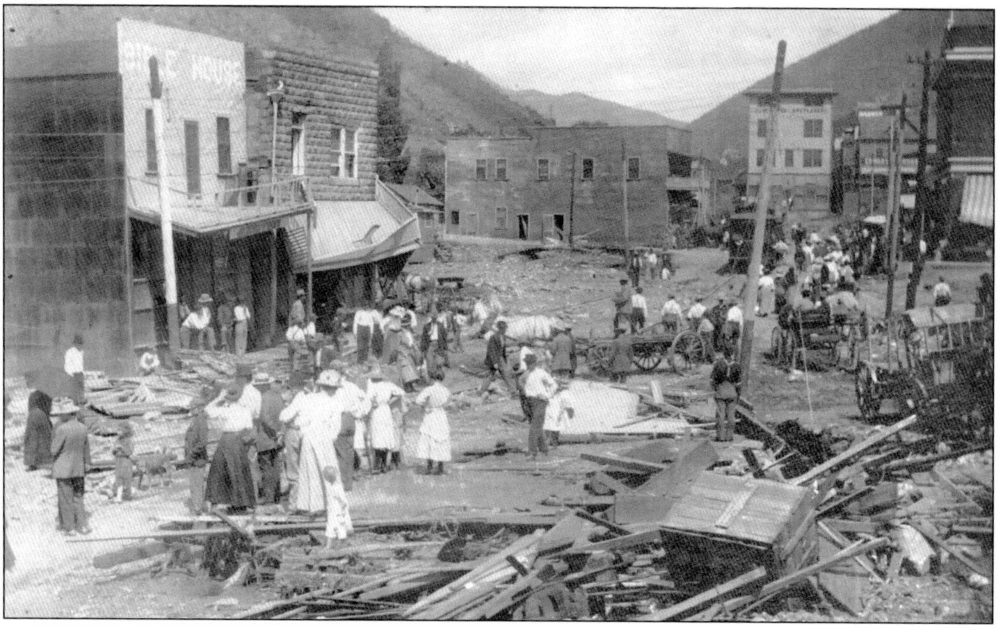

The community of Johnson Addition stood defenseless against the inundation, and the next day, residents took to the streets to survey the ruins. The left side of the above card shows the Bible House that "was rocking around" and became "three feet out of plumb" but managed to survive the ordeal. The building next to the Bible House also remained standing, but its second-level porch was all but gone. The wooden wreckage in the foreground reveals another structure that was not as lucky. Below, the image captured the state of the streetcar tracks after the storm. The ground below them washed away leaving the sinking and undulating tracks behind.

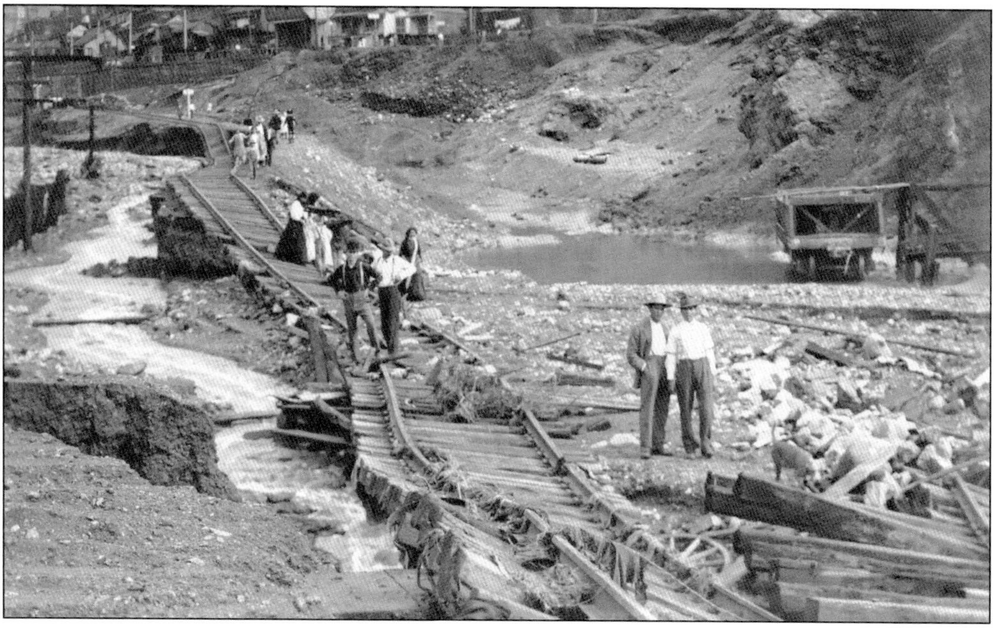

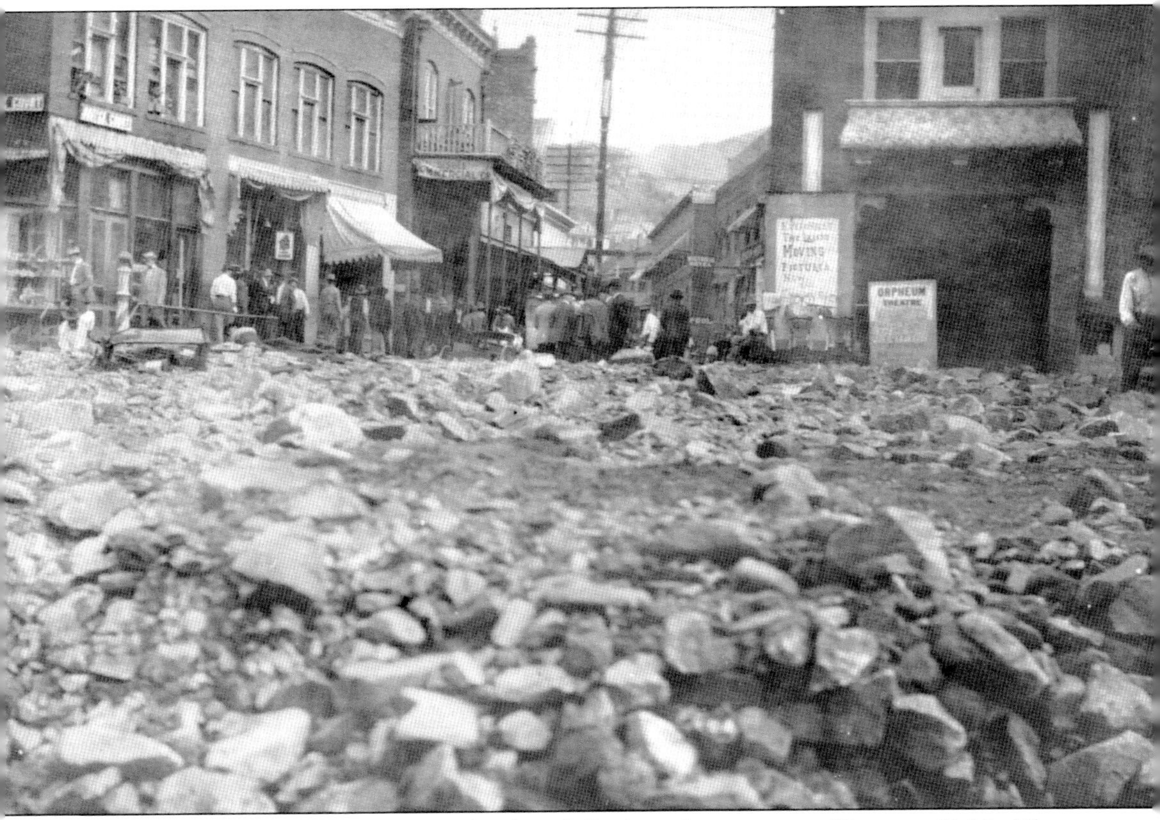

The flood washed thousands of tons of rock and dirt into the entrance of Brewery Gulch. The *Bisbee Daily Review* described the material as having "a distance of more than a block [and] a depth of several feet." Crews worked continuously to open the frequently used street and completed clearing the area in 10 days. Employees of the Warren-Bisbee Railway removed the debris from Brewery Gulch. The company agreed to carry out the task in order to obtain the material so they could use it to fill in washed out areas of the streetcar tracks. By the time merchants added up their losses of inventory, streets were repaired, and homes were rebuilt, the damages of the flood totaled $150,000.

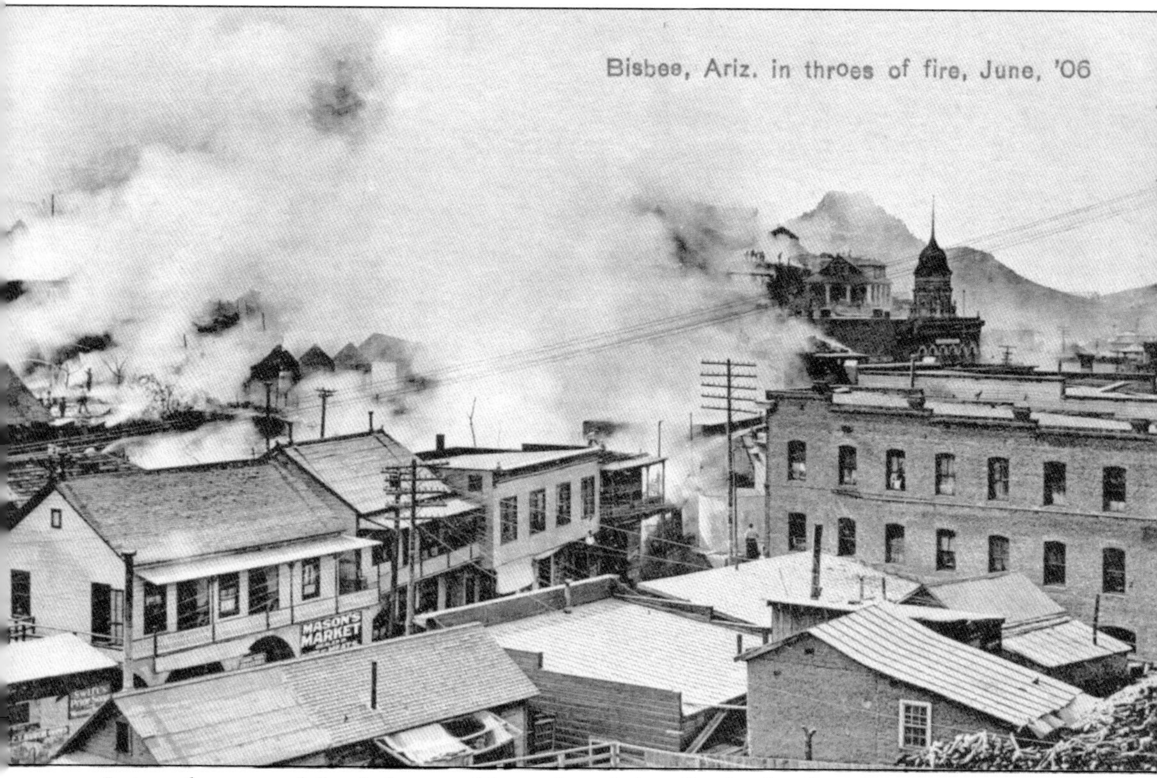

Just as the seasonal floods impacted how the mining district developed, so too did the fires. Two fires stand out as being the most devastating to the area. The first was the Chihuahua Hill fire of 1907, shown here in a postcard that incorrectly dates the event in 1906. Located on the east side of Brewery Gulch, Chihuahua Hill, with its proximity to the area mines and business district, became a well-populated area. In classic Bisbee fashion, the houses there were stacked closely upon the hills, one directly behind the other. This created a perfect environment for a fire to spread, and one did on June 29, 1907. The fire started at the intersection of OK Street and Brewery Avenue and quickly moved up Chihuahua Hill where the "congested condition of the houses made an easy path" for the blaze. The gravity of the impending disaster quickly became evident to onlookers, and "every foot of [fire] hose was brought into action on the hill."

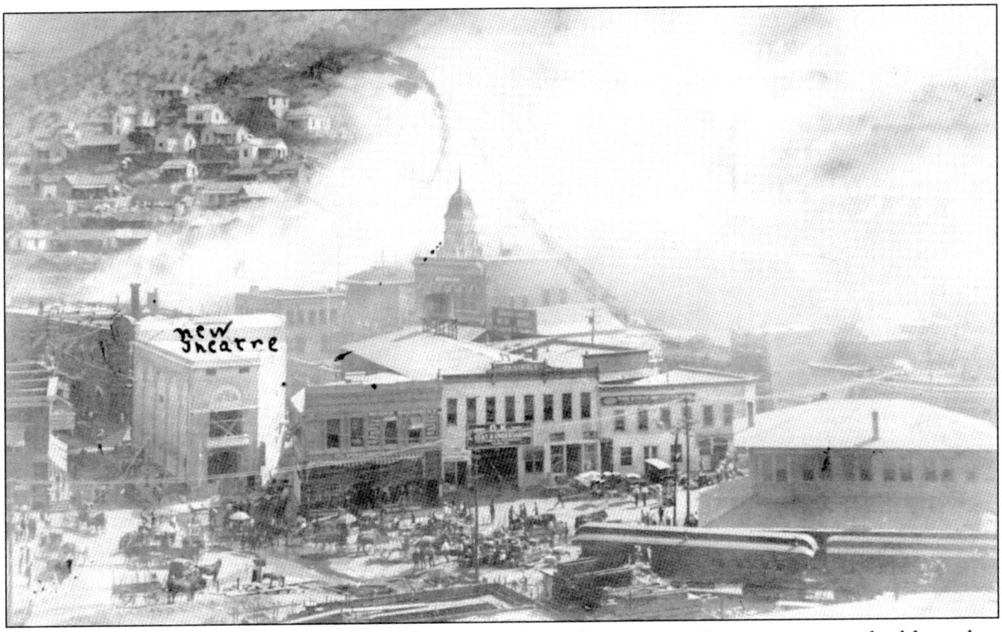

Residents stepped into action and formed bucket brigades, attempting to suppress the blaze that encroached upon dwellings. Soon, it became evident that the firemen needed assistance battling the blaze, and men were ordered out of the mines to fight the flames. The miners also brought dynamite with them to help put out the fire. The *Bisbee Daily Review* described, "Experienced miners went a short distance ahead of the burning section," and proceeded to blow up areas, thus forming a gap that retarded the fire's ability to spread. This tactic, in combination with the work of the firemen, brought an end to the disaster nearly three hours after the fire started. The next day, the local paper broke the news that was already evident to citizens by looking upon the hill that 100 families were now homeless.

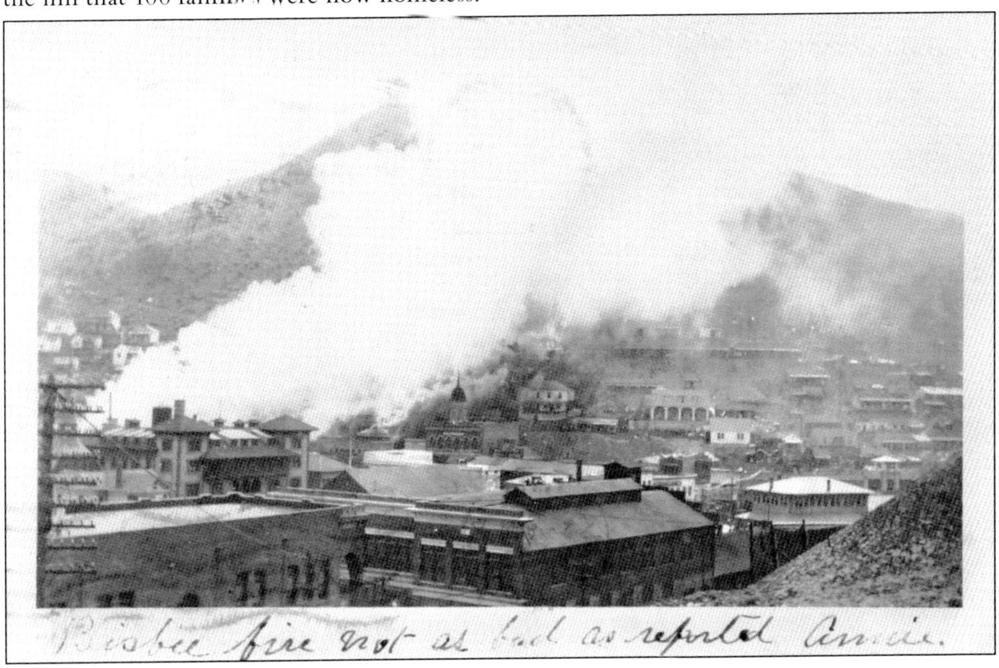

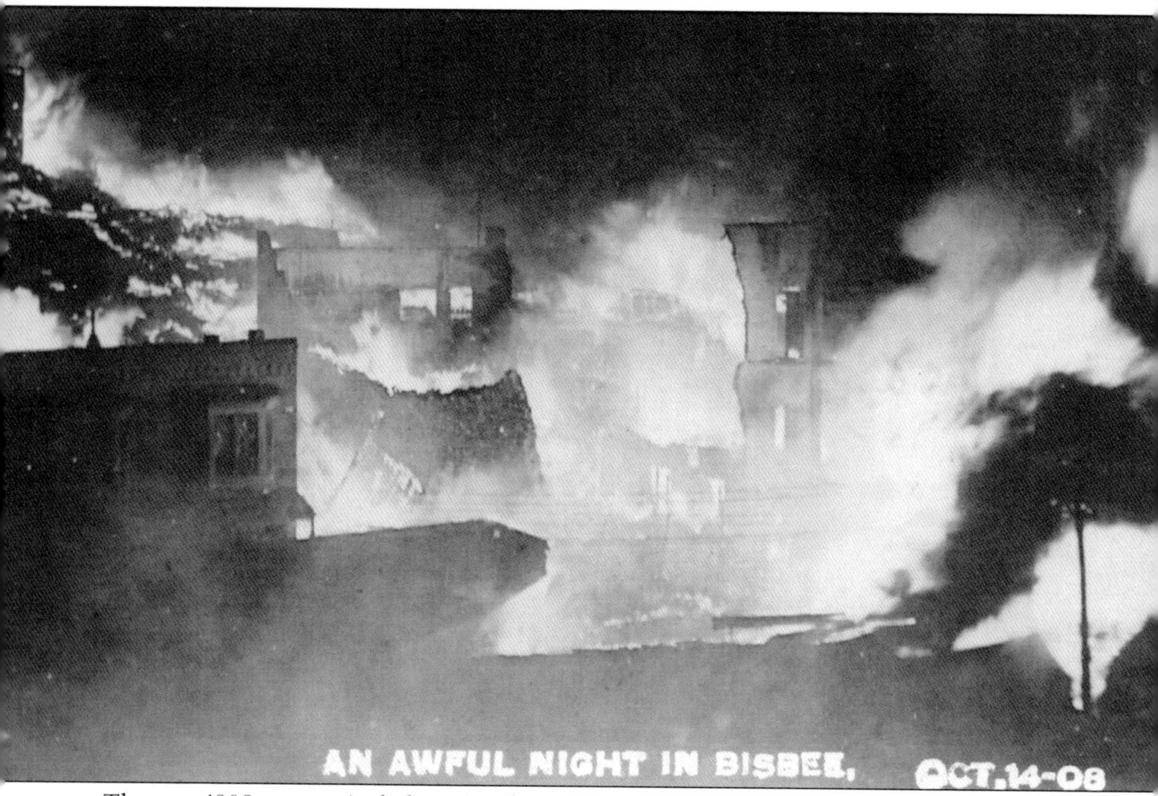

The year 1908 was particularly trying for residents, as three floods had already wreaked havoc upon Bisbee that summer, leaving the mining camp in shambles. However, nothing could compare to what individuals would witness on the evening of October 14, 1908. That night, the worst disaster in the city's history occurred; a fire consumed a large number of homes and a significant portion of the commercial district. The fire is reported to have started in a closet at the Grand Hotel and was seen first by Tom Blair and Sam Frankenberg, one of the owners of the Fair Store. The inferno soon engulfed the building, and occupants made desperate attempts to escape the structure. This included Tony Estes, who the *Bisbee Daily Review* said, "Was forced to jump from the third story of the Grand Hotel into a blanket . . . to save himself from the flames."

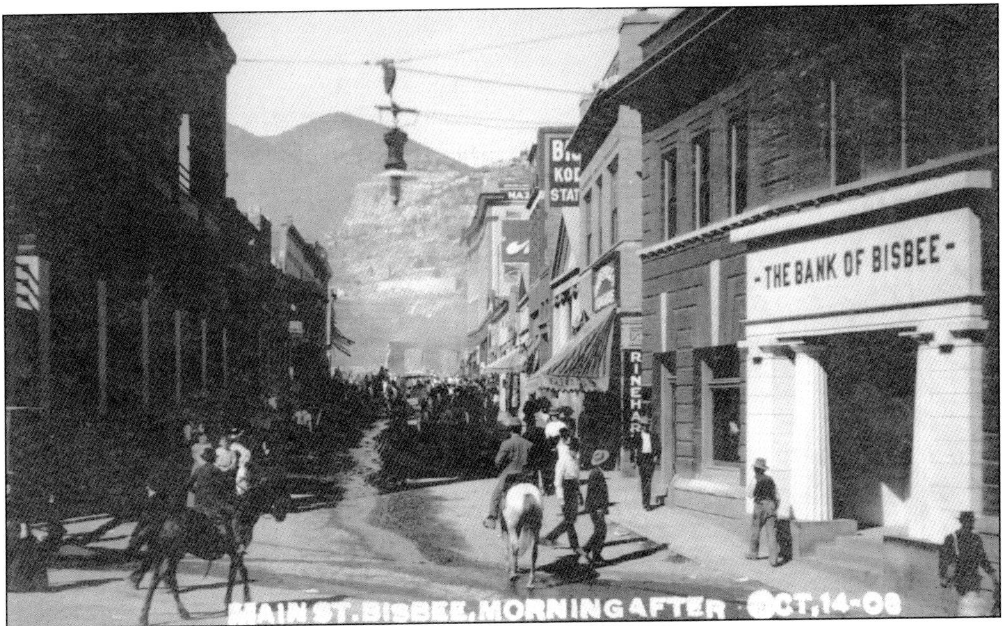

As the fire devoured the Grand Hotel, it began to spread west, up the commercial district on Main Street and north to homes on Clawson Hill. The streets of Bisbee soon filled with residents, some fleeing for their lives with their families. Several stood in sheer disbelief as they watched their worldly possessions burn before their eyes while others courageously risked their own lives to save those in danger. The *Bisbee Daily Review* stated, "Men who went through the tremendous cataclysm at San Francisco . . . pleaded with authorities . . . to dynamite several buildings surrounding the seething center of the furnace," to impede its growth. After hours of arguing, officials finally permitted blowing up the structures to save the city. The method proved successful, and the fire was eventually extinguished.

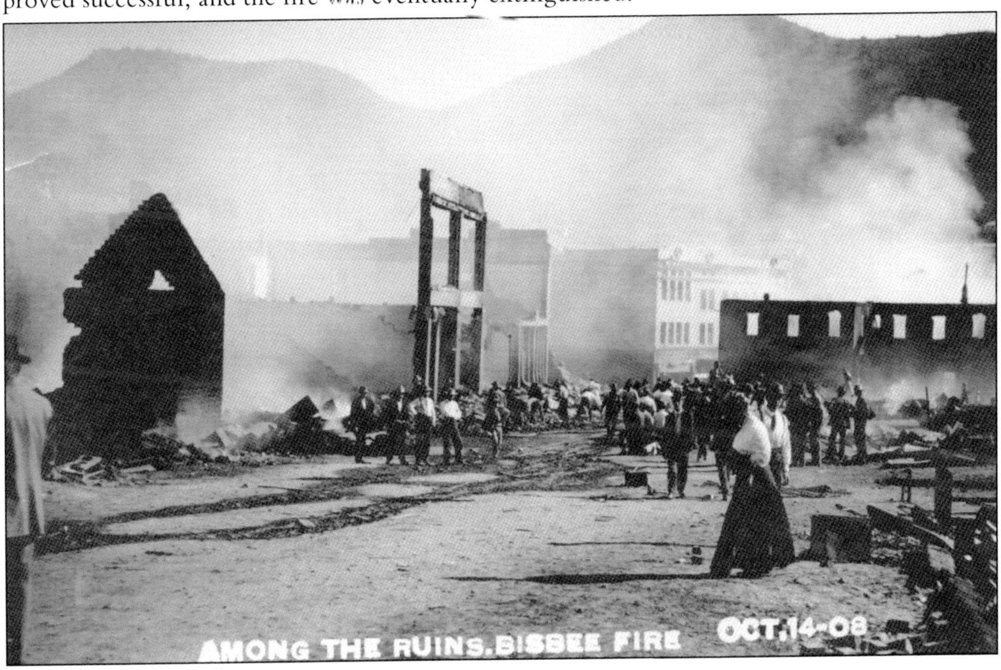

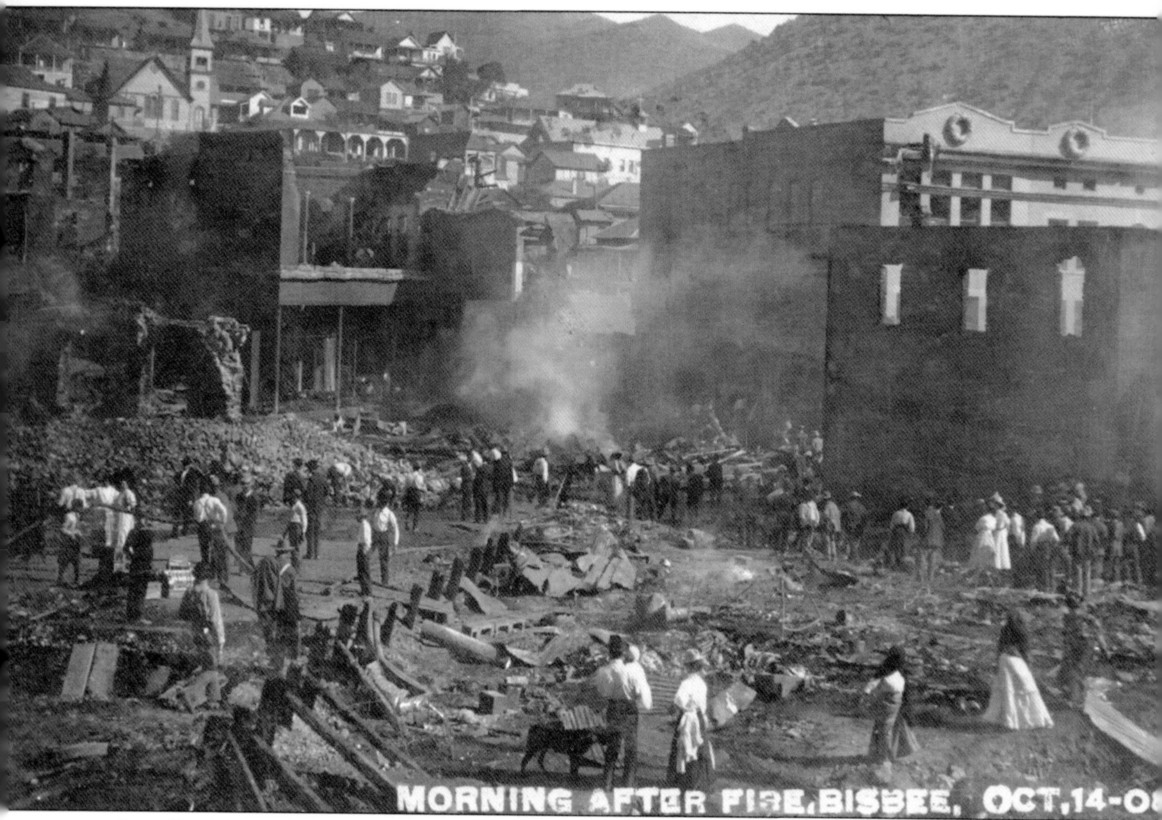

The toll of the fire was devastating. "Every business on Main Street from the Johnson-Henniger building to the north side and Ball & Bledsoe on the south side, and every residence between there and Castle Rock were swept away." The estimated damage from the three-hour fire was larger than Bisbee's previous calamites combined, totaling $750,000. The next day, this image was taken capturing the smoldering ruins of Main Street. The left side of the postcard is where the fire started, amongst the remains of the Grand Hotel.

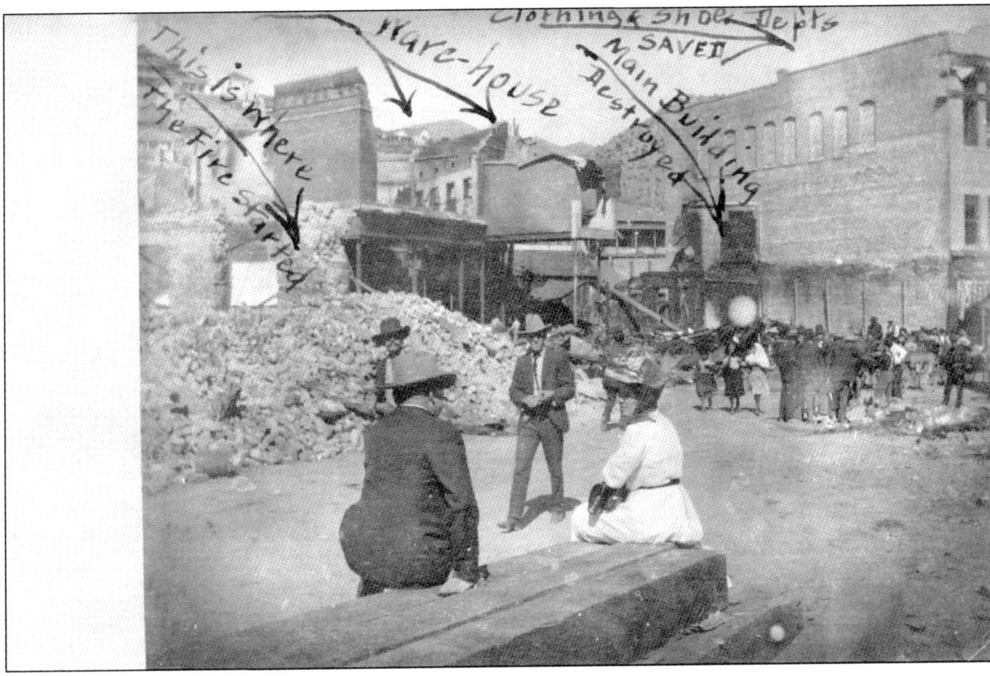

The business that sustained the greatest financial loss was the Fair Store at a total of $300,000. A venture of the Frankenberg brothers, Sam and Ben, the Fair Store's main building was a loss, as well as the warehouse and stock inside of the structures. On the above postcard sent by Ben Frankenberg to a cousin in Virginia he wrote, "[The] section showing our buildings and how they now stand." On the front of the card, he noted the buildings that survived the blaze and those that did not. The image below shows a close up view of the damage sustained by one of their buildings.

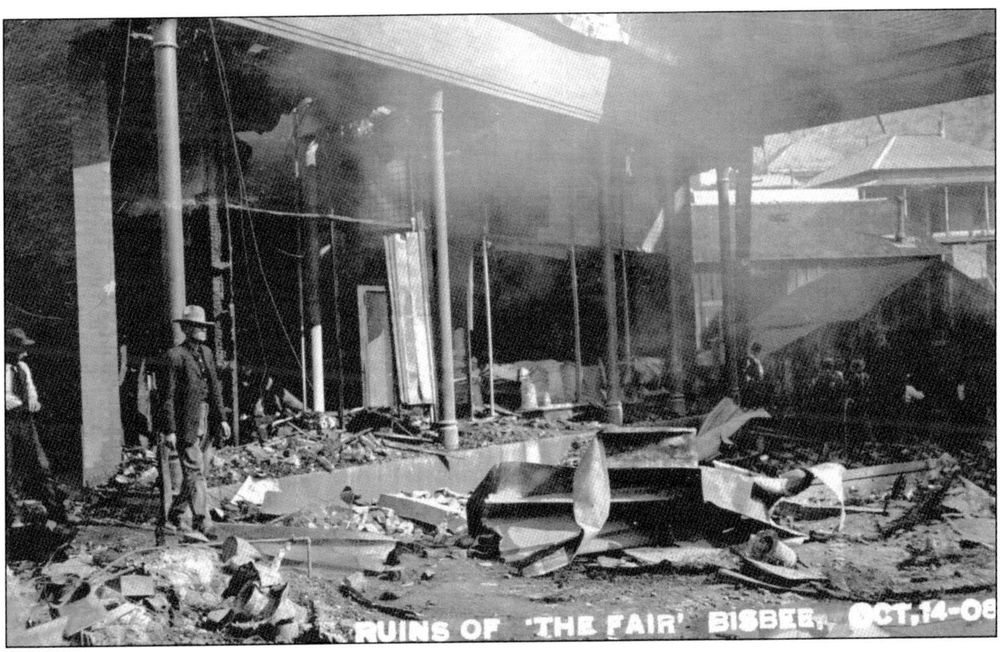

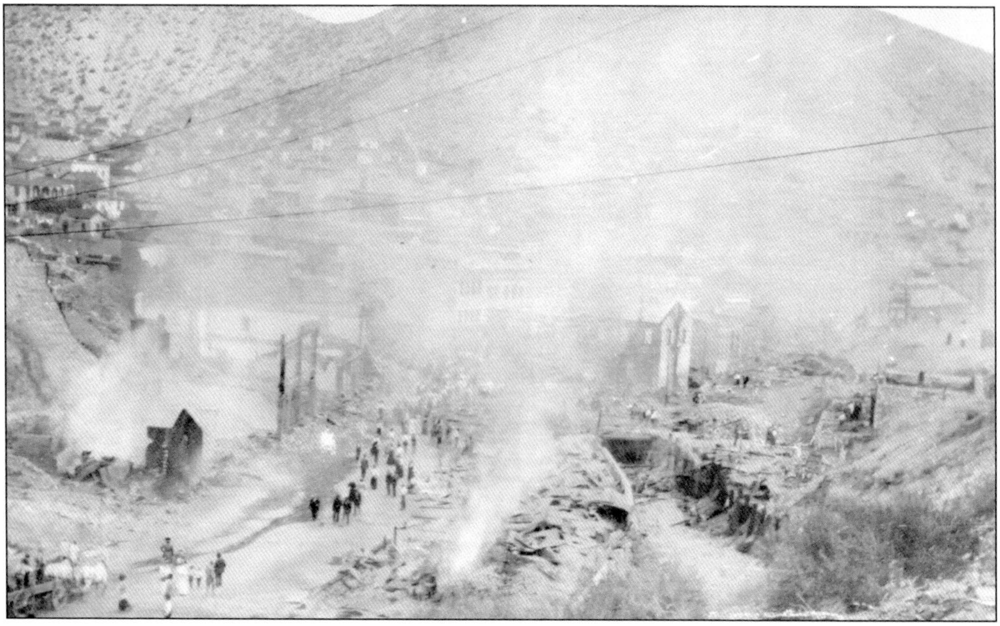

The image above was taken near Castle Rock and reveals the destroyed upper end of Main Street. The merchants on the thoroughfare were determined to rebuild, and within months, new structures began to appear. However, having learned from the tragic experience, this time the buildings were made of more substantial and lasting materials. Despite the considerable loss that impacted the commercial district, an even greater loss was that of the 500 Bisbee residents who lost their homes. Nearly every home on Clawson Hill disappeared as the area was "swept as clean as a prairie" by the flames. The postcard below captures the majority of the surviving structures in the neighborhood. Note that the house on the left used wet blankets on its roof to help prevent embers from lighting the building on fire.

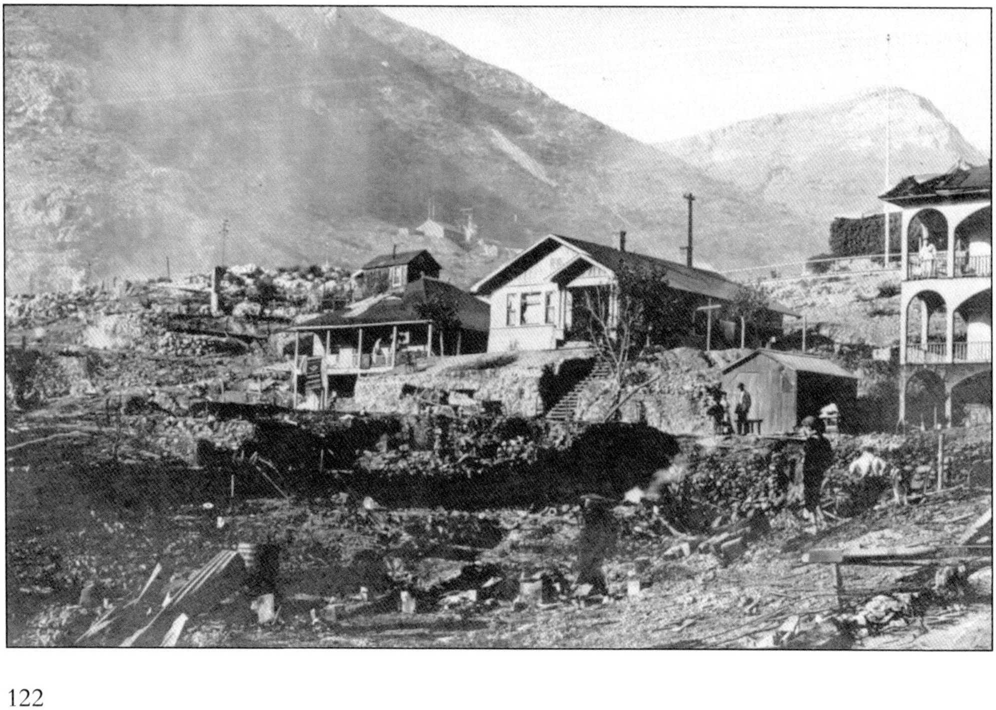

Six

Strife within the Empire

During July 1917, Bisbee was the scene of a dramatic episode in US labor history, as striking miners and union sympathizers were forcibly removed from their homes, placed on a train, and forbidden to return. The troubles began in June when a union, the Industrial Workers of the World, presented the mining companies with a list of demands. Issues such as employees' wages, working conditions, and possible discriminatory practices by the mining company were addressed in the document but were quickly refused by the companies' management.

An infamous organization, the Industrial Workers of the World was well known for its socialist and antiwar viewpoints, such as supporting slowdowns and sabotage within the workplace. The group undoubtedly gained some notoriety during the period by allowing minorities to become a part of their ranks. In fact, the organization was recognized for being the first union in Bisbee that actively sought workers of Mexican descent to join the group.

Rumors accused the antiwar union of intentionally calling this and other walkouts to reduce the essential supply of copper needed to fight World War I. The Industrial Workers of the World was also reported to be backed with German money and have Mexican revolutionaries mixed among the Bisbee strikers. During 1917, these accusations were distressing, largely because of the contents of the Zimmerman telegram. Intercepted by British intelligence months earlier and contrived to distract the United States from the war in Europe, the Zimmerman telegram proposed an alliance between Germany and Mexico. Included in this partnership, Germany offered Mexico "generous financial support" to "reconquer [its] lost territory in Texas, New Mexico, and Arizona."

By July 11, tension grew within Bisbee, as rumors of potential violence by the strikers filtered throughout the city. A meeting was held that night that included Cochise County sheriff Harry Wheeler and Calumet & Arizona manager John Greenway to discuss how to resolve the issues at the mining camp. During the meeting, it was agreed that the strikers would be forcibly gathered and loaded on a train bound for Columbus, New Mexico, the next day.

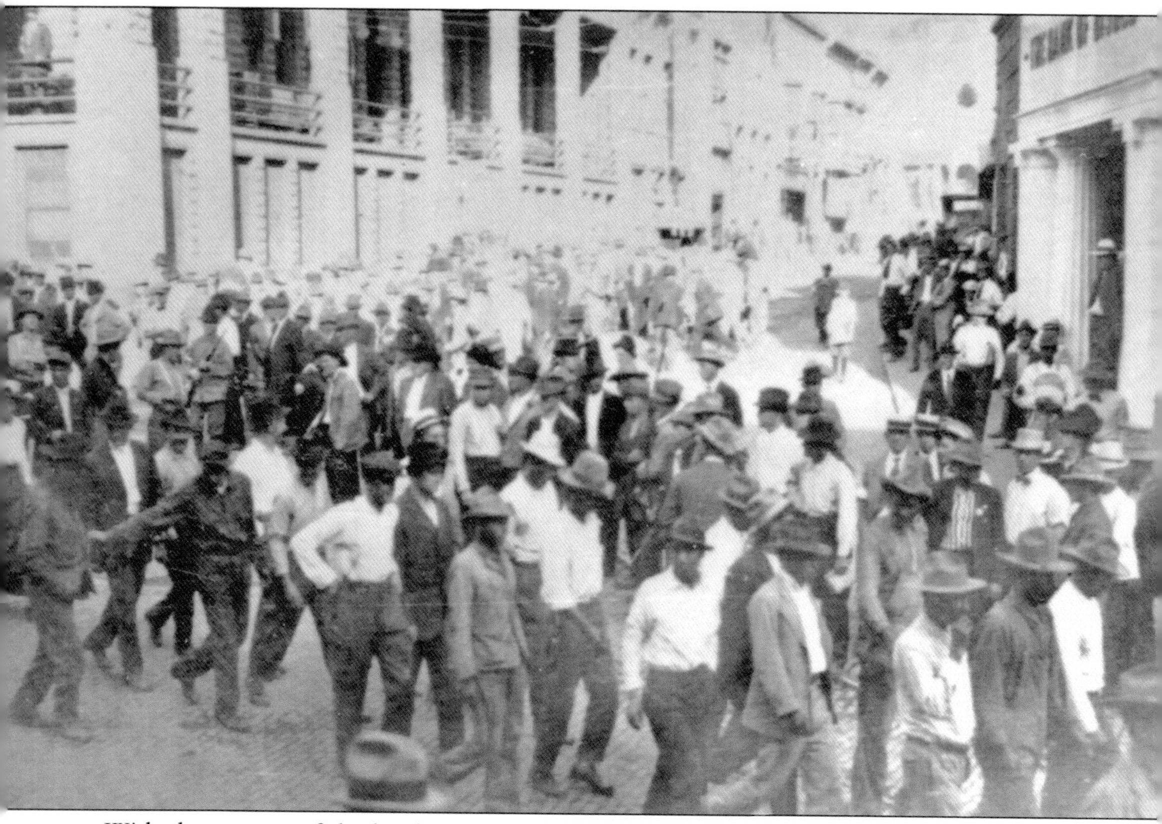

With the support of the local mining companies, Sheriff Wheeler began to put a plan into action. It was decided that the strikers would be arrested "on charges of vagrancy, of treason and of disturbing the peace in Cochise County." Calls were made to posse members in the Bisbee and Douglas areas, and approximately 2,000 armed men entered the mining camp, waiting for Sheriff Wheeler's command to begin the roundup of the strikers. The headline of the July 12, 1917, *Bisbee Daily Review* warned, "ALL WOMAN AND CHILDREN KEEP OFF STREETS TODAY," and as the paper hit the streets of the city at 6:30 a.m., so did the posse. The gathering commenced quickly and within several minutes, 500 union men and sympathizers were forced into the heart of downtown Bisbee. (Courtesy of the Bisbee Mining & Historical Museum, Brophy Collection.)

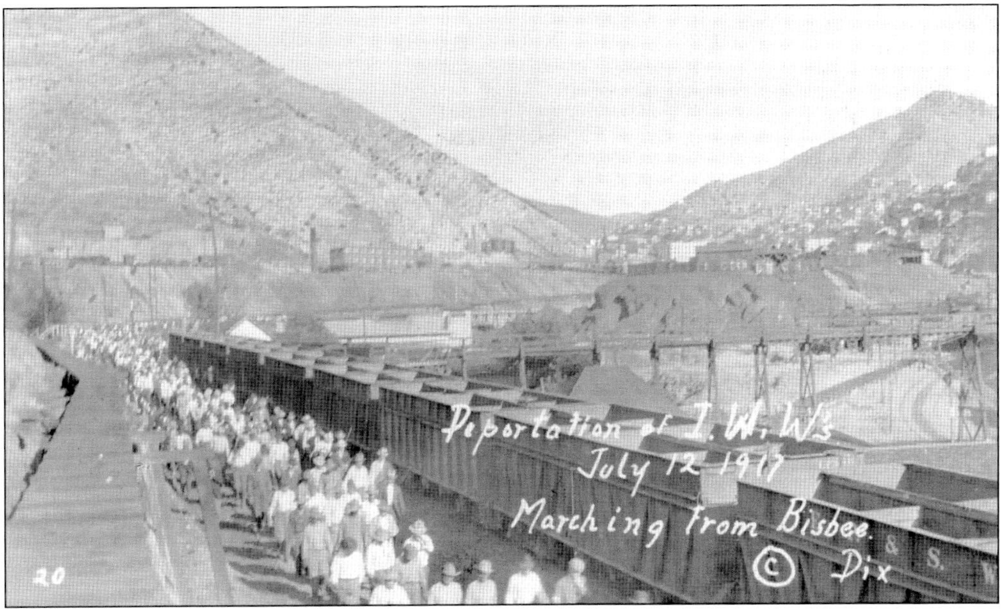

By 7:30 a.m., an immense number of people were gathered downtown. It was decided the crowd would be led to the Warren Ballpark, approximately two miles away, where the men would be loaded on a New Mexico–bound train. These postcards depict the men as they made the journey to the ballpark. As the procession of prisoners walked to Warren, more people were added to the group. In an effort to keep the crowd under control, the captives were kept under armed guard with posse members wearing white armbands to identify themselves.

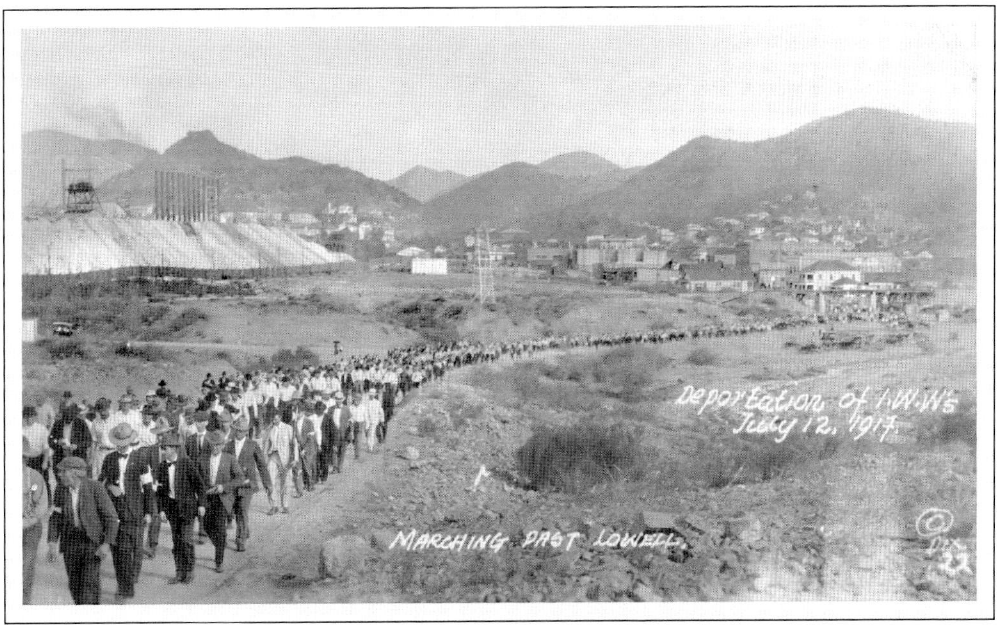

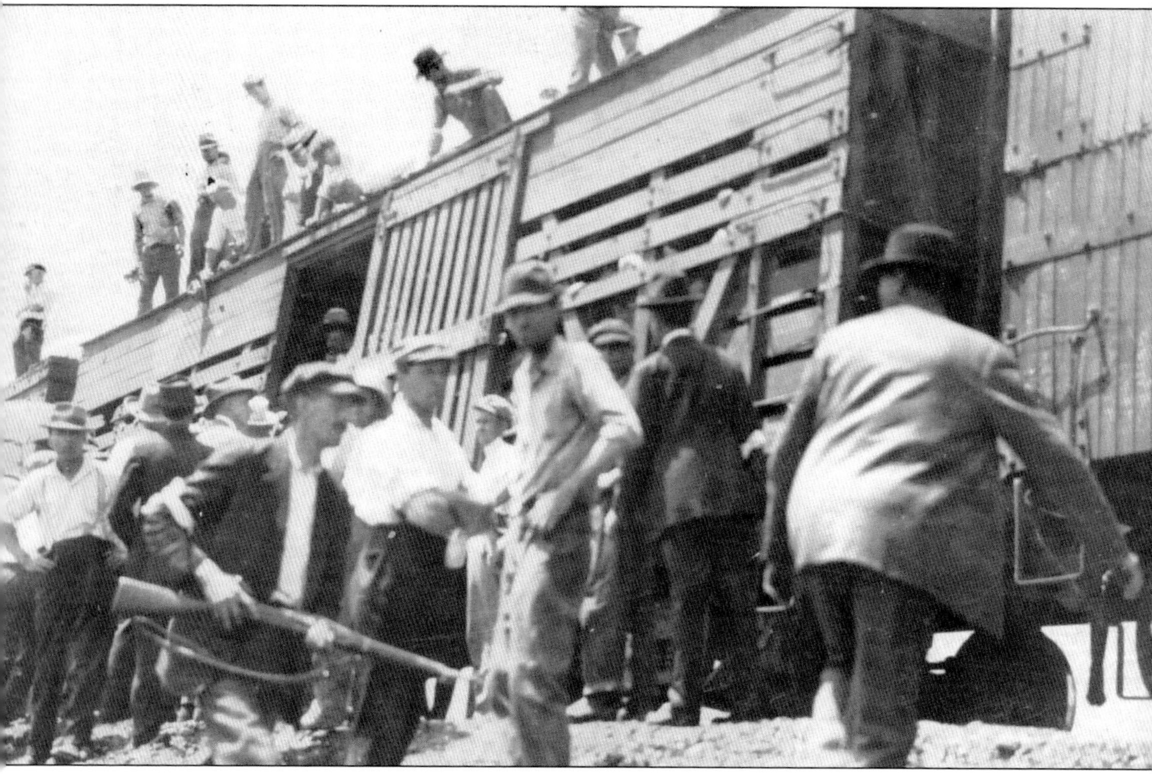

Once at the ballpark, the men were asked if they would go back to work. Those who responded positively were allowed to return home, and those who replied no were loaded into one of the train's cattle cars. Of the 2,000 men who were rounded up that summer morning, 1,186 departed on the train. The men ultimately ended up in the border community of Columbus, New Mexico, where US military officials cared for them. After the deportation, Bisbee citizens took measures to insure strikers would not infiltrate their community. Vigilantes protected all roads leading to Bisbee and railroad points around the clock. To enter the town one was required to provide guards with identification, be questioned intently, and have their vehicle searched. Individuals who appeared suspicious to the vigilantes were ordered to leave the area.

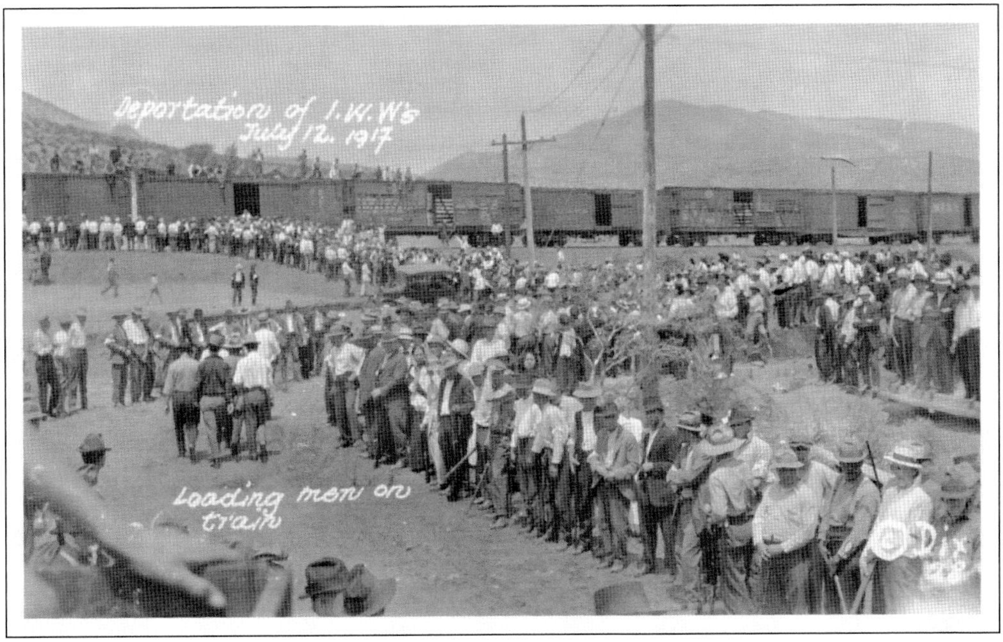

Although the deportation was a clear violation of human rights, no one was found guilty of forcefully removing the men from the community. Kidnapping charges were ultimately filed against 210 posse members who rounded up and deported the men. However, a not guilty verdict was rendered at the trial of the first defendant. The court ruled Bisbeeites acted within their rights under the Law of Necessity, which stated the community had a right to protect itself from a threat, whether it was real or perceived. Believing the other indictments would result in the same verdict, the charges against the remaining posse members were dropped.

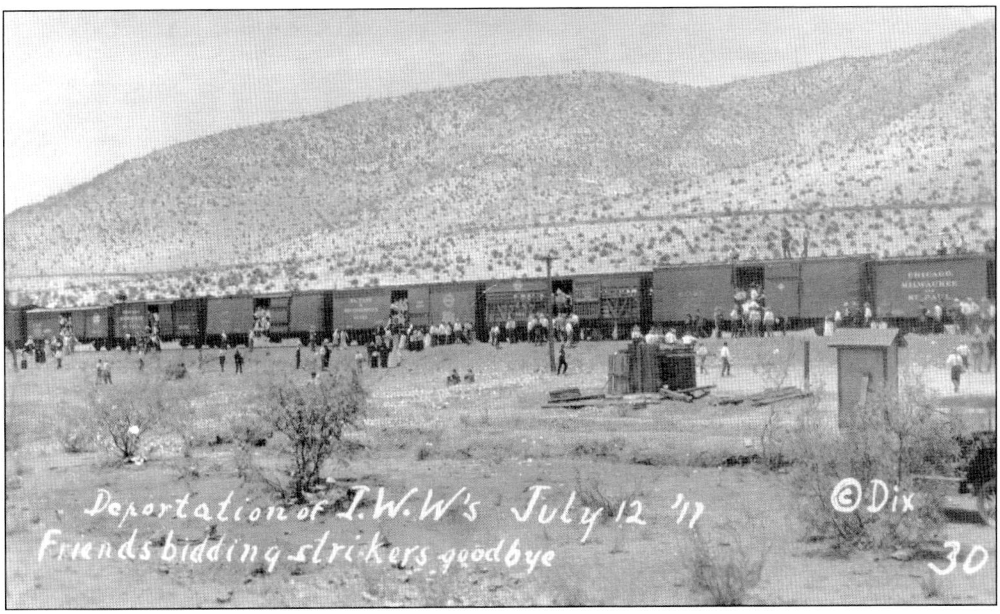

Discover Thousands of Local History Books Featuring Millions of Vintage Images

Arcadia Publishing, the leading local history publisher in the United States, is committed to making history accessible and meaningful through publishing books that celebrate and preserve the heritage of America's people and places.

Find more books like this at
www.arcadiapublishing.com

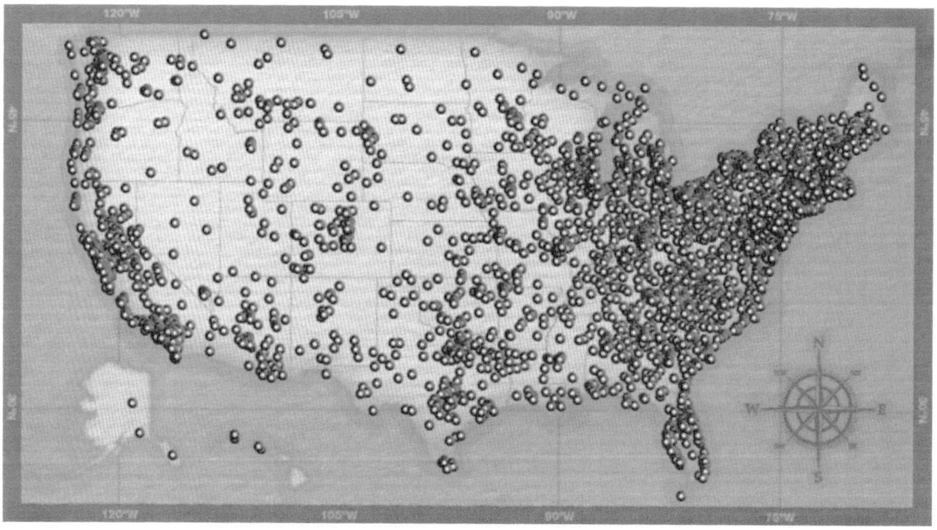

Search for your hometown history, your old stomping grounds, and even your favorite sports team.

Consistent with our mission to preserve history on a local level, this book was printed in South Carolina on American-made paper and manufactured entirely in the United States. Products carrying the accredited Forest Stewardship Council (FSC) label are printed on 100 percent FSC-certified paper.

MADE IN THE USA